THE ANDERTON BOAT LIFT

The electrically operated boat lift circa 1950, with serveral Weaver flats visible in the Anderton Basin beyond. Note the staircase up to the aqueduct and control cabin, which has since been dismantled, and the footbridge across the entrance to the lift, also since removed. In comparison with earlier views of the lift and basin, it can be seen that all of the salt chutes have been removed.
Courtesy The National Waterways Museum, Gloucester

THE ANDERTON BOAT LIFT

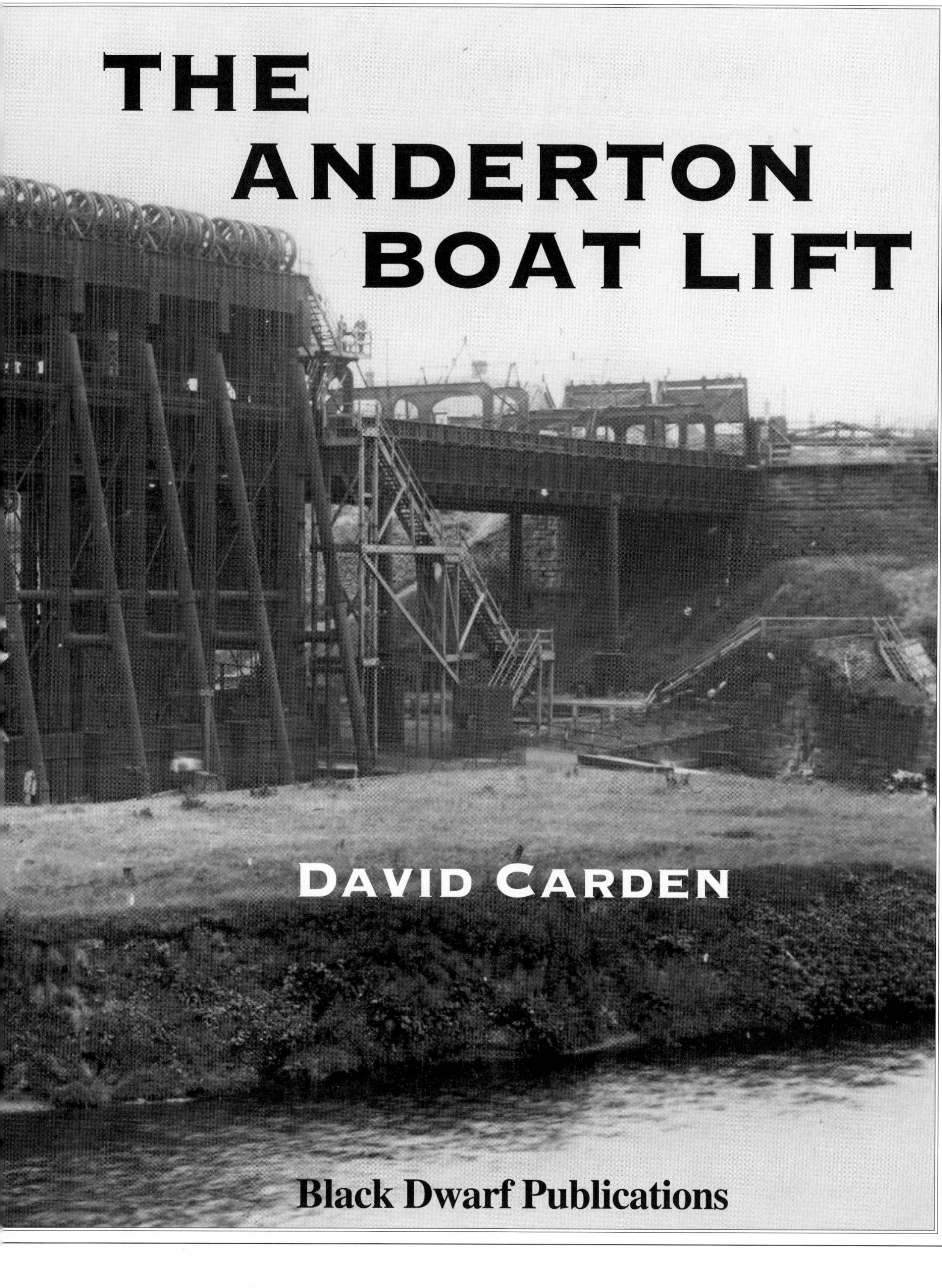

DAVID CARDEN

Black Dwarf Publications

An overall view of the lift in 1965, showing its approach aqueduct, with derelict land previously occupied by warehouses in the foreground and a small motor coaster in the Anderton Basin in the background.
Courtesy The National Waterways Museum, Gloucester

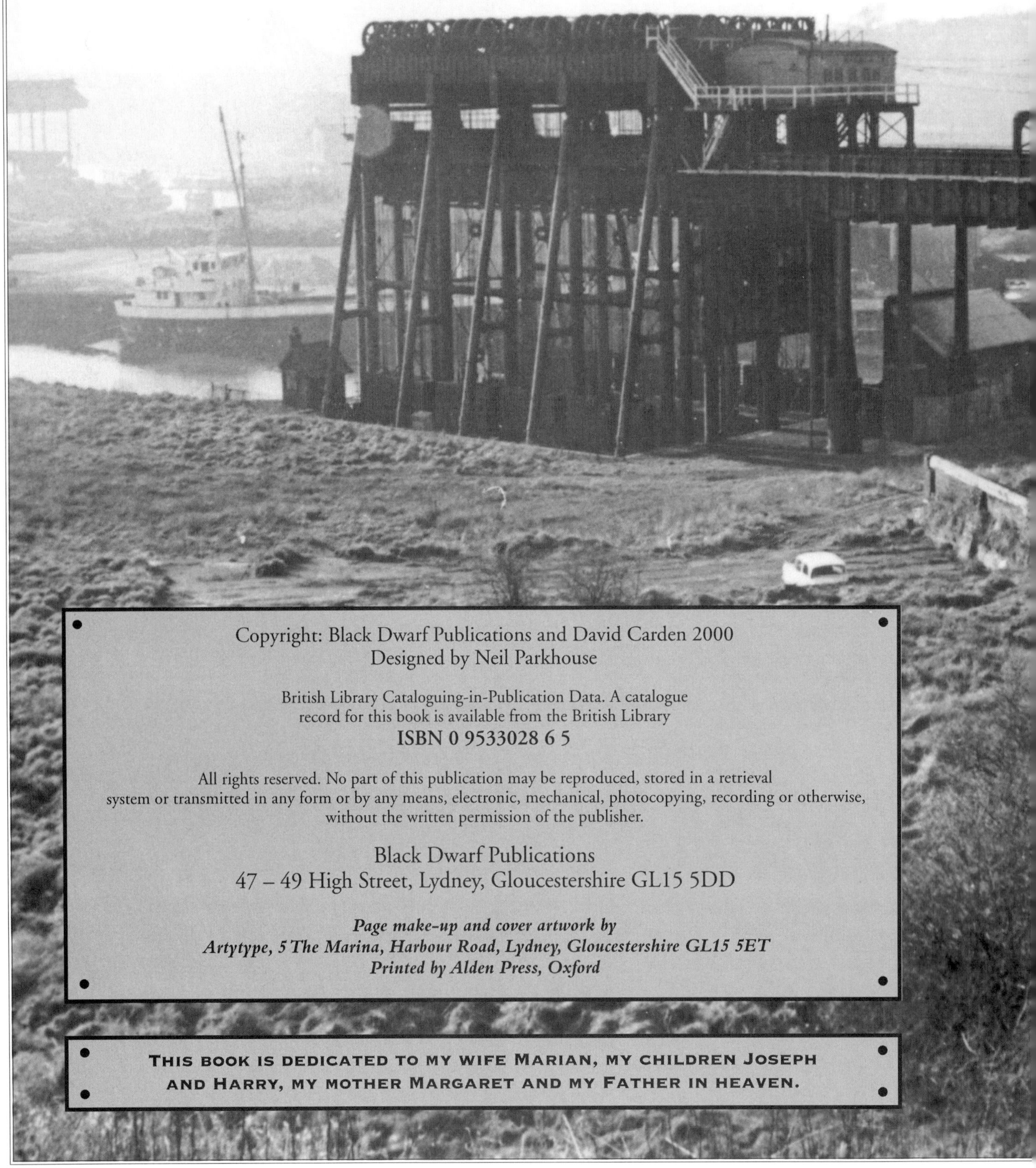

Designed by Neil Parkhouse

British Library Cataloguing-in-Publication Data. A catalogue record for this book is available from the British Library

ISBN 0 9533028 6 5

Black Dwarf Publications
47 – 49 High Street, Lydney, Gloucestershire GL15 5DD

Page make-up and cover artwork by
Artytype, 5 The Marina, Harbour Road, Lydney, Gloucestershire GL15 5ET
Printed by Alden Press, Oxford

THIS BOOK IS DEDICATED TO MY WIFE MARIAN, MY CHILDREN JOSEPH AND HARRY, MY MOTHER MARGARET AND MY FATHER IN HEAVEN.

THE ANDERTON BOAT LIFT

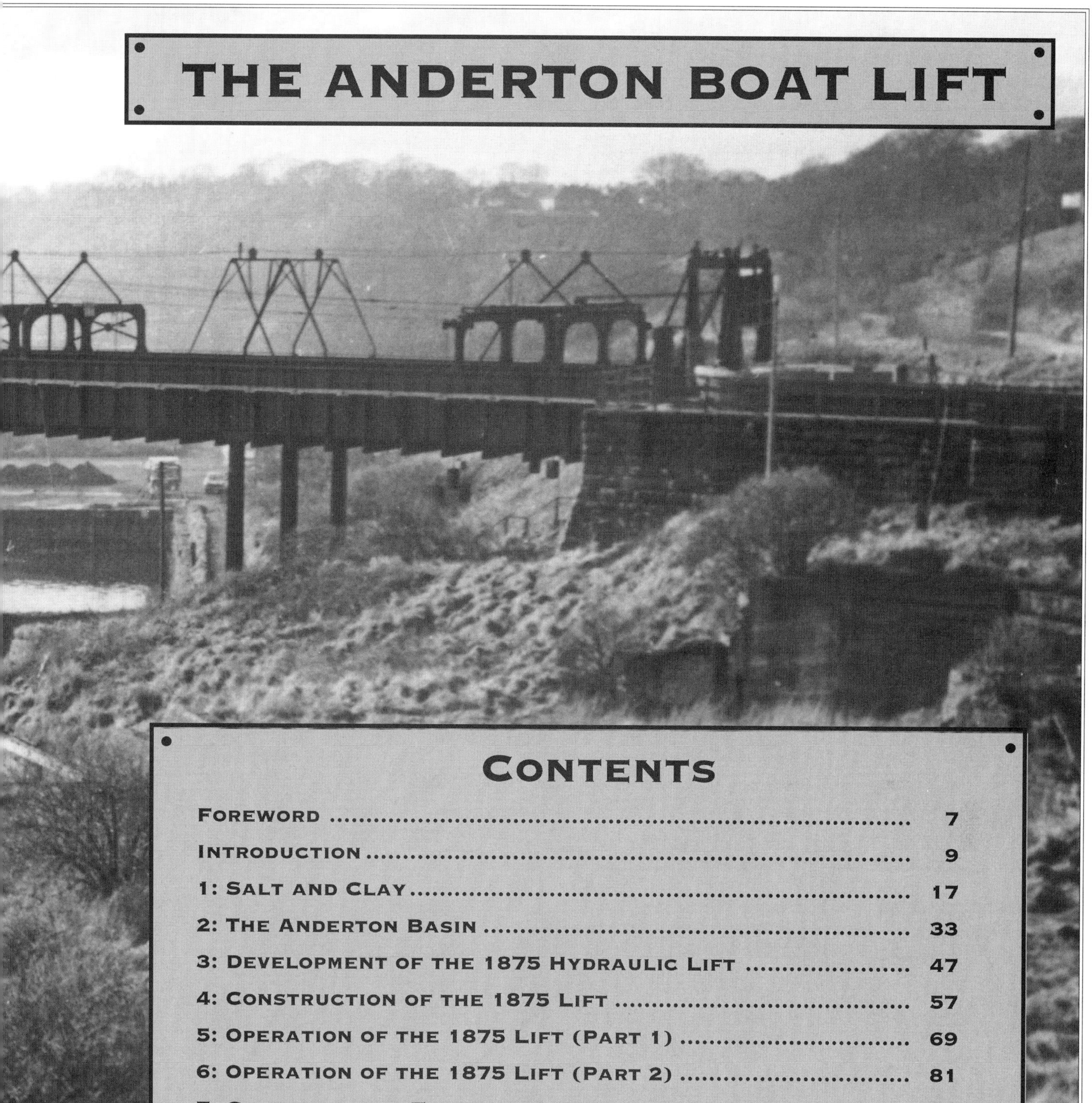

Contents

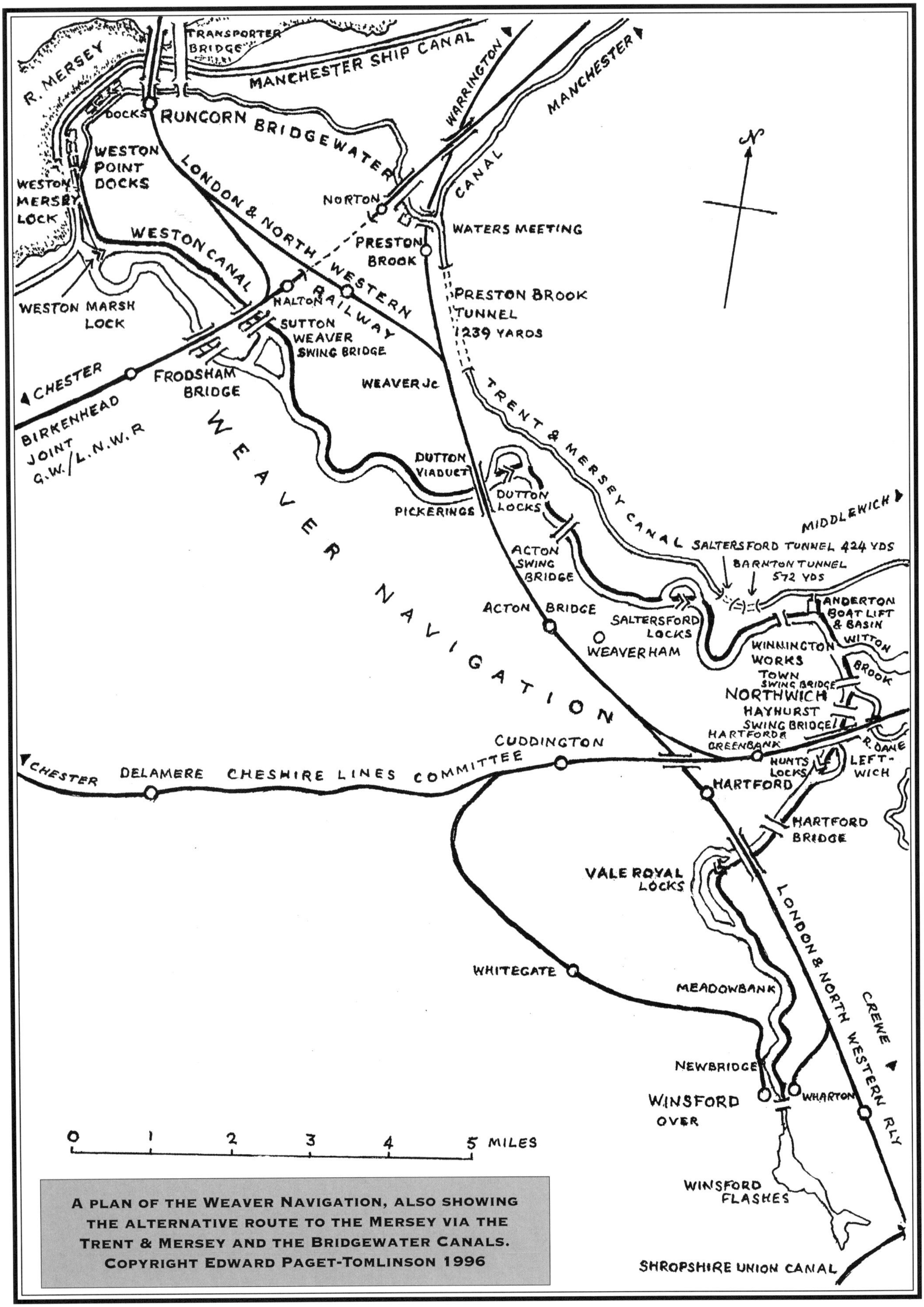

A PLAN OF THE WEAVER NAVIGATION, ALSO SHOWING THE ALTERNATIVE ROUTE TO THE MERSEY VIA THE TRENT & MERSEY AND THE BRIDGEWATER CANALS. COPYRIGHT EDWARD PAGET-TOMLINSON 1996

Foreword

The Anderton Boat Lift is one of those structures which can truly be described as inspirational. Its development spawned the construction of hydraulic boat lifts in Europe and Canada; even the conversion of the Lift at the beginning of the century was ingeniously carried out and its method of operation has been adopted for the world's most modern boat lift at Strépy-Thieu in Belgium. Telling the story therefore is important not just as a piece of history but as a description of how human endeavour and inspiration can overcome seemingly insurmountable problems.

David Carden tells this story not only from the perspective of a qualified engineer but from the viewpoint of somebody enchanted by the structure and with knowledge of the key role that individuals have had in its fascinating history.

These twin themes not only chronicle events but identify a parallel with events today. The vision shown by the Weaver Navigation Trustees is mirrored by The Anderton Boat Lift Trust and now The Waterways Trust in working with British Waterways to find a solution to restoring this large moving metal structure.

There is much still to be done before the Anderton Boat Lift can be operational once again but the story he tells about the Lift and the people involved will stimulate and galvanise us all, by giving us the confidence that the pathway we are taking to this goal has been trodden before.

Derek Cochrane
Regional Director British Waterways

The official seal of the Weaver Navigation Trustees, 1721-1895.
Courtesy The Boat Museum, Ellesmere Port

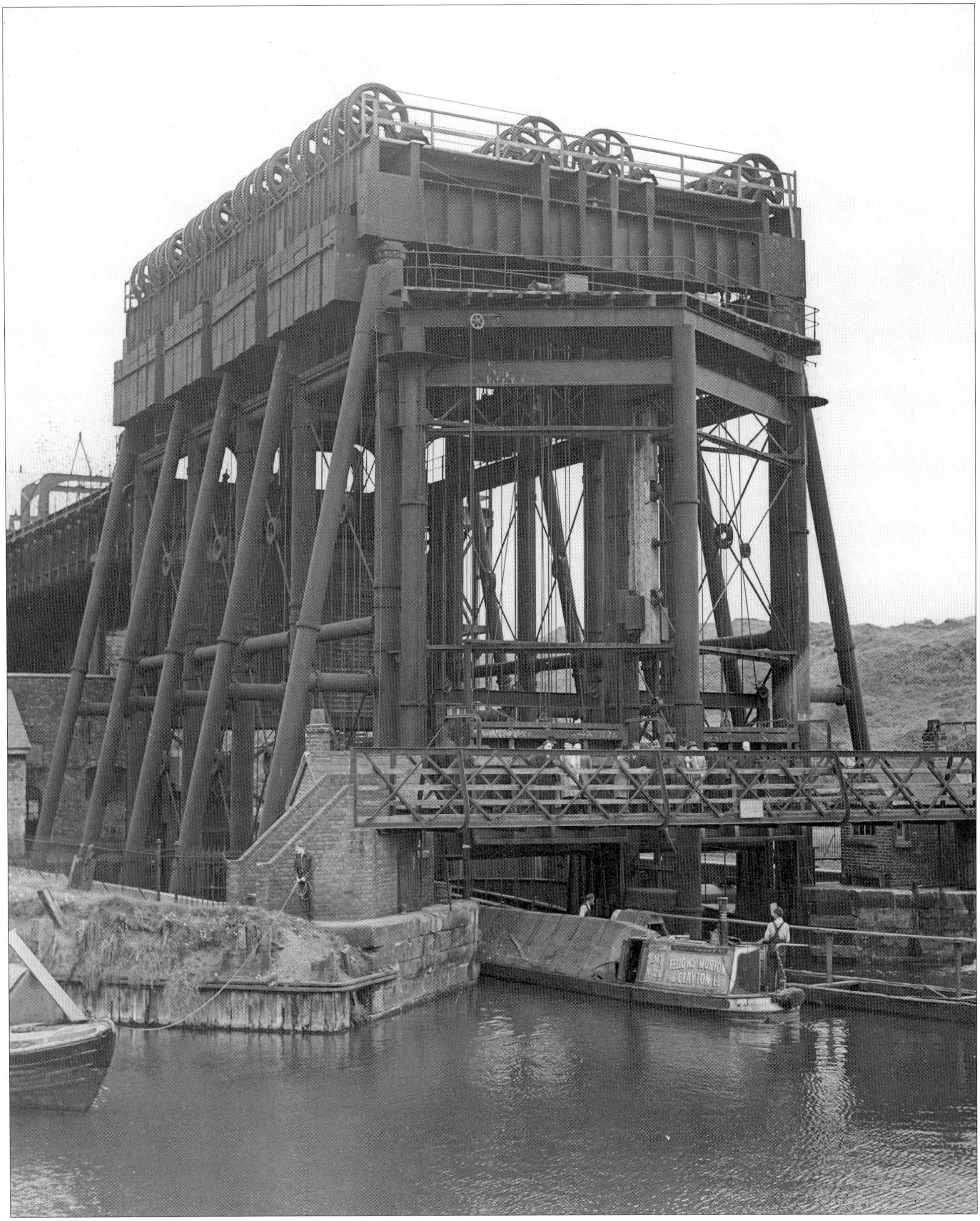

Following Nationalisation of Britain's transport systems in 1948, the Weaver Navigation, including the Anderton Boat Lift, and the Trent & Mersey Canal came under the jurisdiction of the British Transport Commission. This photograph was taken in summer 1948 whilst the lift was being visited by Sir Cyril Hurcomb (Chairman of the British Transport Commission 1947-53) and other dignataries. The visit to the lift was part of a tour of transport related sites in Cheshire. Beneath the visitors on the footbridge across the river entrance to the lift, a Fellows, Morton & Clayton boatman guides his narrowboat into the western caisson in order to ascend the lift to the canal 50 feet above. As an interesting aside, this is the very photograph on which the well known painting on the cover of the March 1949 edition of Meccano Magazine *was based (see rear dust jacket).* Neil Parkhouse collection

INTRODUCTION

'REBUILDING A MONSTER'

'I was dismayed to read in your excellent article on the Anderton boat lift, that English Heritage has offered half a million pounds of taxpayers' money towards restoring the structure to a working state.

I have no quarrel with the principle of restoration – I admire those hard working people who restore steam locomotives – but I question the need to rebuild what is arguably the country's ugliest structure.

Current issues such as the status of engineers will be set back a pace or two if we persist in rebuilding this monstrosity – is this really the face of engineering we wish to present to society?'

Letter by G. Davies printed in the *New Civil Engineer* on 7 March 1996

Well, I suppose there is only one correct reaction to a letter like this, which is that we are all entitled to our own opinion. Who knows, there may be many people in this country who, if they knew about English Heritage's intentions, would share Mr. Davies' views. To be frank, until a few years ago, I may well have been one of them.

Indeed, five years ago, I'm not sure if I had even heard of the Anderton Boat Lift and, if I had, I certainly would not have had any views about the rights and wrongs of its restoration. At that stage, I might have had some sympathy with the suggestion that restoration works should be left to enthusiastic amateurs, although one can't help thinking that restoring a structure of the size of the Anderton Boat Lift would dampen the enthusiasm of the most ardent amateur.

Since 1994, however, I have been fortunate enough to have become professionally involved in the project to restore the Anderton Boat Lift. My current employer Lewin, Fryer & Partners, an engineering consultancy based at Hampton in Middlesex, was appointed by British Waterways - North West Region to undertake detailed investigations with the ultimate objective of restoring the boat lift and returning it to operational duty. One of my responsibilities, alongside my colleagues, has been to organise and manage site surveys to identify the present condition of the 120 year old structure. During the course of these, I have, either by necessity or fortuitously, come across a number of historical documents relating to the lift and have met with several individuals who have had direct experience of the operation of the lift before it was closed on safety grounds in 1983. By such means, my curiosity in the history of the lift was aroused and I started on the long road of inquiry that has culminated in this book. As a consequence, I am now an enthusiastic devotee of the Anderton Boat Lift and am totally convinced that it should be restored to full working condition. I must now confess that I am no longer able to discern whether it is a thing of beauty or ugliness.

During the research for this book I have acquired an admiration for the work of those Victorian engineers, namely Edward Leader Williams, Edwin Clark and John Watt Sandeman, who, between them, conceived the idea of an hydraulic boat lift at Anderton and managed to persuade the Trustees of the Weaver Navigation to proceed with its construction in 1875. Similarly, I admire the efforts of Col. J. A Saner who, on behalf of the Trustees, was responsible for the conversion of the boat lift to electric power in 1908.

There are surely only a few who would deny that the original 1875 hydraulic lift was an engineering masterpiece, elegant both in design and operation. This is admirably stated in Vale Royal Borough Council's recent publicity brochure about the lift:

'In consultation with Edwin Clark, a prominent civil engineer, he, (Edward Leader Williams), produced a design that was a unique and magnificent example of the Victorian's mastery of cast iron and hydraulics . . . Completed in 1875, graceful in appearance, simple in use, and above all efficient, the Lift was hailed as a marvel of the era.'

I agree wholeheartedly with the sentiments of this extract, although I would take issue with its accuracy in attributing the design to Leader Williams. My understanding was that Leader Williams conceived the idea of a boat lift at Anderton, but Edwin Clark was responsible for the actual design of the hydraulic lift.

It must be said that, in contrast, the structure that remained after the conversion of the lift to electrical operation in 1908 was far from elegant. In a recent article in *The Times* newspaper the lift is described, rather unflatteringly as:

'*. . . an early North Sea oilrig rising from the riverbank.*'

In its defence, however, the 1908 electrical conversion construction was a masterly feat of practical engineering that resulted in an entirely new boat lift being erected over the badly corroded hydraulic lift, with only minimal disruption to the heavy traffic between the River Weaver and the Trent & Mersey Canal. Therefore, in my humble

The hydraulic boat lift, circa 1880, with various warehouses and salt chutes in the background and materials stacked in the foreground. Also clearly visible is the original stairway between the upper platform of the lift and the island, which was replaced during conversion to electrical operation in 1908. Steve Astles collection

opinion, this structure, as well as the 1875 hydraulic lift, should be given a place of importance in Britain's engineering heritage

So, for the sake of balance, I include below a copy of a letter from Mr. Phil Parker, Project Manager for British Waterways - North West Region, which was published in the New Civil Engineer in April 1996:

DUTY OF CARE

'I was somewhat disappointed to read G. Davies' letter (NCE 7 March) about the Anderton Lift.

Beauty is to be seen in the eye of the beholder; steam trains are dirty, uneconomical to run and inefficient, but I would not question restoration. The lift was, and is, one of the most innovative structures the world has seen and surely that alone makes preservation a necessity.

The current status of engineers is a burning question. What price do we pay if we let our civil engineering roots and heritage disappear? Is this the image we wish to present to the outside world that we do not care, nor do we have the ingenuity or the will to restore structures, such as the boat lift.

I think our profession does care, even to save the 'ugliest structure in the world'.'

★★★★★★★★★★

Sadly, if you visit Anderton today all you will see is a massive rusting structure with a confusion of columns, beams, diagonal struts, ties and other ironwork elements. The lift has been inoperable since 1983 and is stripped of the heavy cast iron winding gear and counterweights that belonged to the 1908 electrical conversion. They are now stockpiled in rows on the ground alongside. The discovery of severe corrosion in many of the structural elements led to the enforced closure of the boat lift on safety grounds. And so, the Anderton Basin, which 100 years ago was bustling with commercial activity, now has an air of dereliction, with the public excluded by means of padlocked gates and high security fencing. The site is reverting back to nature with trees and shrubs taking over at ground level and the birds using the higher parts of the corroded structure as a secure place to bring up their nestlings. Under such circumstances, it is difficult for a visitor today to visualise how the lift and site would have been in its heyday at the end of the 19th century. Then, the Anderton Basin was a major crossroads in Britain's inland transport system and was bustling with activity. Raw materials for the Staffordshire Potteries, such as clay, flint and coal, were carried up to the Trent & Mersey Canal from the River Weaver, whilst salt and finished potteryware were carried down. A sense of the commercial importance of the lift site is conveyed even in this description of the lift in the 1950's by Tom Foxon:

'Almost every morning we would awake to find a handful of Knobsticks, piled high with crates and casks of pottery ware, waiting to go down the Lift when it opened at 8am. The next traffic would be the boats which had started up from Middlewich that morning, some of them empty, having discharged coal at Seddon's, but the Mersey, Weaver boats were loaded with gravel chippings for Trafford Park. The morning trade from Weston and Runcorn would appear from about 10 am onwards with a variety of cargoes such as china clay, china stone, flints and felspar. Around 1pm we would see the first traffic from Manchester, consisting mainly of

flour-laden Mersey, Weavers and Coventry bound Cowpars.'
Extract from *Anderton for Orders* by Tom Foxon

'Knobstick' was a nickname given to a narrowboat from the Anderton Company Limited. The origin of this was seemingly one of the company's Boat Marshals, who was well-known for brandishing a stick with a prominent silver knob.

The contrast with the apparently abandoned site that exists today is dramatic. However, behind the scenes there is much activity. British Waterways, with the backing of English Heritage, are committed to restoring Edwin Clark's 1875 hydraulic lift to full operating condition with the structural components of Saner's 1908 electrical conversion being retained in situ as a static monument. British Waterways' commitment was well demonstrated by Laurence Morgan, who was their Project Manager in 1994, when he was quoted on the front cover of the October 1994 edition of *Newways* (an internal British Waterways' magazine), as saying:

'What excites me about the lift is being involved in its 1990s restoration – it has already been restored once, in 1908. I feel that I'm contributing to its next 90 years.'

Laurence has now moved on to pastures new but I very much share his enthusiasm. Like him, I feel privileged that, by some quirk of fate, I have found myself involved in the restoration of the historic boat lift. It may well be that, in years to come, I will be taking my grandchildren to Northwich, as I have already done on several occasions with my children, to talk to them about my small role in the continuing history of the Anderton Boat Lift.

This book is my humble attempt to tell the story of the Anderton Boat Lift. To my knowledge no other recent book exists on this subject and I hope that the product of my efforts will adequately fill this void.

★ ★ ★ ★ ★ ★ ★ ★ ★ ★

Before starting on the story of the lift itself, I feel that some explanations are needed.

Firstly to give sufficient background, I have chosen to begin telling the story many years before even the conception of the lift, dealing with the origins of salt mining in Cheshire and pottery-making in Staffordshire. These commercial activities were the main driving forces behind the rapid industrialisation of the north west of England in the 18th and 19th centuries, which eventually led to demands for improved trading links with the outside world. In the course of time, these demands resulted, firstly, in the canalisation of the River Weaver, then the construction of the Trent & Mersey Canal and eventually the building of the Anderton Boat Lift.

Subsequent chapters deal with the development of the Anderton Basin, the design, construction and operation of the hydraulic lift, its conversion to electricity in 1908 and the operation of the electrical lift through the two world wars until its closure in 1983. In the penultimate chapter I have described the studies and investigations

The electrically powered boat lift in operational condition circa 1965, with security fencing around the structure. In comparison with the previous view, it can be seen that the electrical conversion works completed in 1908 altered the appearance of the lift quite dramatically.
Colin Lynch collection

A number of working narrowboats waiting in the canal basin at the top of the electrical lift in the 1930s, with the Winnington chemical works in the background. A loaded pair appear to be waiting to descend, on the right, with a single boat to their left. All are sheeted up, indicating they are carrying perishable cargoes – possibly pottery – and all three are horse boats, as indicated by the towing pins protruding through the top plank and sheets. The empty boats could well be at Anderton 'for orders'. Note the gate on the eastern side of the aqueduct is partly drawn.
Courtesy The Boat Museum, Ellesmere Port

which have been carried out in the years following the closure of the lift, whilst in the final chapter I have touched upon the proposed restoration of the structure.

During the writing of this book I consider that my role has been more one of a collector and collator of facts rather than that of an historian. Let me explain. As a consulting engineer working on behalf of British Waterways, I was in the fortunate position of having direct access to several historical records concerning the lift. These whetted my appetite and, without too much effort, I was able to gain access to yet more records from other sources. The content and quality of these records was such that there was little need for interpretation or imagination; what was needed tended to be sifting and sorting. This process, which has taken me a year or two to complete, has been very much a labour of love, which started with a natural curiosity into the origins of the Anderton Lift but gradually developed into a passion that my wife and children would probably describe as an obsession.

With respect to presentation, I make no apologies for the frequent use of quotations whether from historical records or other documents. The quotations have been selected principally to give the reader an impression of the atmosphere of the times but also because earlier writers often described events in such a way that I would find difficult to better.

Similarly, I make no apologies for the relatively large number of photographs and diagrams scattered through the pages. I am aware of their value to the reader in providing a welcome distraction from the text, which inevitably, at times, can be rather weighty. Speaking for myself, I am often daunted by page after page of the written word. I also make no apologies for the large selection of photographs of the 1906-8 reconstruction of the lift, courtesy of a marvellous archive held at the Boat Museum, Ellesmere Port. Not only was the reconstruction sufficiently interesting in both engineering and visual terms to merit their inclusion, they also serve to balance somewhat the positive dearth of pictures which have so far surfaced of the construction of the original lift in 1875; only one view appears in these pages.

All but the last chapter of the book are historical and have been prepared principally by reference to primary sources (e.g. Minutes of Proceedings of the Weaver Navigation Trustees), but also by use of secondary sources. I have given careful consideration as to how best to give suitable credits within the book to the various sources of data that I have made use of. I rejected the idea of giving references within the body of the text as it can be an unwelcome distraction to smooth reading, and I eventually decided to give references at the end of each chapter. I hope the reader will not be offended by this.

I have deliberately chosen to keep the last chapter of the book relatively succinct. It was not easy to decide at what point in time the book should end. I eventually concluded that it would be very appropriate to end with the announcement in The Times newspaper of a grant from the Heritage Lottery Fund that will enable the restoration of the lift to be commenced. By so doing, I have left the

An aerial photograph of Winnington and the static lift in 1989, after removal of the headgear; the cast iron wheels can be seen stacked by the river bank, on this side of the boat lift. The Winnington chemical works dominate the left foreground and the village of Barnton appears in the right background. The Weaver Navigation, centre, and the Trent & Mersey Canal, right, run across the view in parallel, into the left distance.

Courtesy Jefferson Air Photography

'A Massive Rusting Structure' – The Boat Lift 1999

Images of the inoperable boat lift in October 1999, described in the text as 'a massive rusting structure, with a confusion of columns, beams, diagonal struts, ties and other ironwork elements.' For safety reasons, the heavy cast iron headgear and counterweights were taken off the weakened structure in November 1987 and stacked at ground level alongside the lift.
Photos Martin Grant and author

telling of the story of the restoration work itself to another day and another book.

Finally, I have no doubt that, as hard as I have tried and as long as the book is, there will still be inaccuracies and areas of insufficient detail. I can only apologise for these at the outset and hope that you do not find any errors too much of a distraction.

★ ★ ★ ★ ★ ★ ★ ★ ★ ★

I wish to express my thanks to all those (alas too many too list) who contributed in some way or other to the writing of this book, including all who have allowed me to make use of their documents or photographs or have bravely agreed to be interviewed by me. I especially wish to thank the following individuals and friends who have offered me particular support, whether in the form of practical assistance, encouragement, or other means:

Steve Astles, Engineering Supervisor at British Waterways - North West Region.
Peter Bentham, Engineering Manager at British Waterways - North West Region.
Margaret Carden, my mother.
Marian Carden, my wife.
David Currie, structural design engineer and colleague at Lewin, Fryer & Partners.
Brian Haskins, Area Engineer for British Waterways at Northwich from 1968 to 1988.
Tony Hirst, Director of the Ellesmere Port Boat Museum (now retired).
Roy Jamieson, The National Waterways Museum, Gloucester.
James Keeley, friend and ex-colleague at Lewin, Fryer & Partners.
Dave Palmer, friend and colleague at Lewin, Fryer & Partners.
Phil Parker, Project Manager at British Waterways - North West Region.
Neil and Dennis Parkhouse, Black Dwarf Publications.
Julie Sharman, Manager of Border Counties Area, British Waterways - North West Region.
Diana Sumner, The Boat Museum, Ellesmere Port.
Jeremy Traynor, friend and photographer.

I would also like to thank **Derek Cochrane**, Regional Director of British Waterways - North West Region, for contributing the Foreword to the book.

I would also like to express my thanks to the various organisations that have assisted me in my endeavours, especially the Boat Museum, Ellesmere Port; British Waterways - North West Region; Lewin, Fryer & Partners, the National Waterways Museum at Gloucester, the Potteries Museum & Art Gallery, Stoke-on-Trent and the Salt Museum at Northwich.

★ ★ ★ ★ ★ ★ ★ ★ ★ ★

Before closing this chapter, I cannot resist giving some quotes about the Anderton Boat Lift that contrast, almost to the point of embarrassment, with the views of Mr. Davies, whose letter was quoted at the very start of the chapter. The following statements have been taken from the bid document to the Heritage Lottery Fund as prepared in April 1998 by British Waterways, in partnership with the Anderton Boat Lift Trust:

'The Anderton Boat Lift is a seminal piece of the physical and historical landscape of Great Britain, and to fully appreciate its status and significance it must be placed within that general context . . .'

'What is remarkable about the Anderton Boat Lift is the sheer scale of the engineering involved in the designing, construction and use of the structure itself. The Anderton Boat Lift was a product of the late Victorian desire to build bigger and better than ever before, allied to the great leap forward in technology afforded by developments in hydraulic engineering, and the industrial revolution in particular . . .'

'The Anderton Boat Lift represents the shift from an industrial revolution driven by exploitation and an impetuous commercial imperative to an environment where scientific, engineering and technological theory and experiment were harnessed to advance what we know as the Second Industrial Revolution.'

Notwithstanding the obvious hyperbole in the above extracts, the Anderton Boat Lift is undoubtedly unique and is regarded, by those who know about such things, as a very important part of Britain's engineering heritage. This is exemplified by the fact that the Anderton Boat Lift is listed by British Waterways as one of 'The Seven Wonders of the Waterways', alongside such famous canal structures as Pontcysyllte Aqueduct on the Llangollen Canal, Barton Swing Aqueduct on the Bridgewater Canal, Foxton Inclined Plane on the Grand Union Canal, Bingley Five Rise staircase locks on the Leeds & Liverpool Canal, Harecastle Tunnel on the Trent & Mersey Canal and the Crofton Steam Beam Engines on the Kennet & Avon Canal.

INFORMATION SOURCES FOR INTRODUCTION

Letter by G. Davies printed in the New Civil Engineer, *7 March 1996.*
Publicity Leaflet entitled The Anderton Lift - A Wonder of the Waterways, *published by Vale Royal Borough Council.*
Article from The Times *newspaper, 31 March 1999.*
Letter by P. Parker printed in the New Civil Engineer, *April 1996.*
Anderton for Orders, *by Tom Foxon. 1988. J.M. Pearson & Associates.*
Article from Newways, *October 1994.*
Bid document to Heritage Lottery Fund, *prepared by British Waterways April 1998, in partnership with Anderton Boat Lift Trust.*
The Seven Wonders of the Waterways, *a promotional video by Video Active for British Waterways.*

The reality of working life in the Cheshire Salt Industry at the beginning of the 20th century. Working conditions were oppressively hot and humid, the hours were long and the work itself was back breaking. This photograph shows the removal of crystallized salt from the pan, known as the 'draught'. The saltmaker draws the salt to the edge of the pan with his rake, whilst his assistant, behind, shovels it into a tub, which stands in the brine. The other two workmen are the 'happer' and his mate. The happer formed the salt into squares by pressing it into moulds. A brief moment's respite has been taken from these tasks – which were normally carried out at speed – for the benefit of the photographer.

Courtesy The Salt Museum, Northwich

CHAPTER 1
SALT AND CLAY

'The opening of the (Trent & Mersey) canal had an immediate effect on the trade of the Weaver at Winsford, which fell from 13,917 tons in 1776 -77 to 5,284 tons in 1779 -80, and meant that raw materials for the Potteries and crates of outgoing pottery were now going by canal instead of by the Weaver and land carriage. On the other hand, the bulk of the Northwich trade remained on the river, coal coming up and salt going down.'
Extract from *The Canals of the West Midlands* by Charles Hadfield

As can be deduced from the above extract, salt mining and pottery making were two industries in North West England which had a major impact on the industrialisation of the region. The Anderton Boat Lift was constructed in 1875 as a commercial response to the costly problem of transshipping goods between two of the waterways in this region, namely the River Weaver Navigation and the Trent & Mersey Canal. Before that, however, the waterways themselves were developed primarily to meet the needs of the salt and pottery industries for better transport facilities, both within the region and to the outside world. In the case of the River Weaver, it was made navigable in 1721, principally due to pressure from salt proprietors in Cheshire and salt merchants in Liverpool, whilst the Trent & Mersey Canal was constructed in 1777 as a direct consequence of the pottery industry in Staffordshire. Both of these industries have, over the centuries, had an immense impact in the region in social, economic and landscape terms.

It would not be possible to tell the story of the Anderton Boat Lift without some reference to salt and clay. The difficulty is knowing just how far back to go in time and the amount of detail needed to give a sufficiently clear picture. So, to be on the safe side, let us start briefly in the Triassic Period.

★★★★★★★★★★

SALT AND THE WEAVER

The currently most popular theory put forward by earth scientists is that about 200 million years ago, during the Triassic Period, that part of the Earth's surface which was eventually to become North West England contained a large but relatively shallow basin, with a connecting channel to the open sea. This basin would, for whatever reason, periodically fill up with seawater and, due to the British Isles at that time having the same latitude as that of the Sahara Desert today, a high rate of evaporation resulted in higher and higher concentrations of salt in the diminishing stock of water. When the resulting mixture in the basin reached critical density, rapid crystallization of the salt took place, with the subsequent formation of rock salt. Numerous repetitions of this cycle, together with complex water currents in and out of the basin over a period of several million years, is thought to have created the abundant deposits of rock salt which now underlie the counties of Cheshire and Shropshire.

Two main rock salt beds have been identified underlying the Cheshire Plain, both being nominally 100 feet thick. The beds are closest to the ground surface at Marbury, a mile or two north of Northwich and close to Anderton, where the upper bed is just 100 feet below ground. More typically, the upper bed is found 200 feet or so below the surface. It has been estimated that, in total, the salt field contains about 400,000 million tonnes of salt.

Where the rock salt is well below the surface, the deposits are largely unaffected by groundwater movements. Closer to the surface, however, the rock salt is dissolved by groundwater which leads to the creation of 'wild' or natural

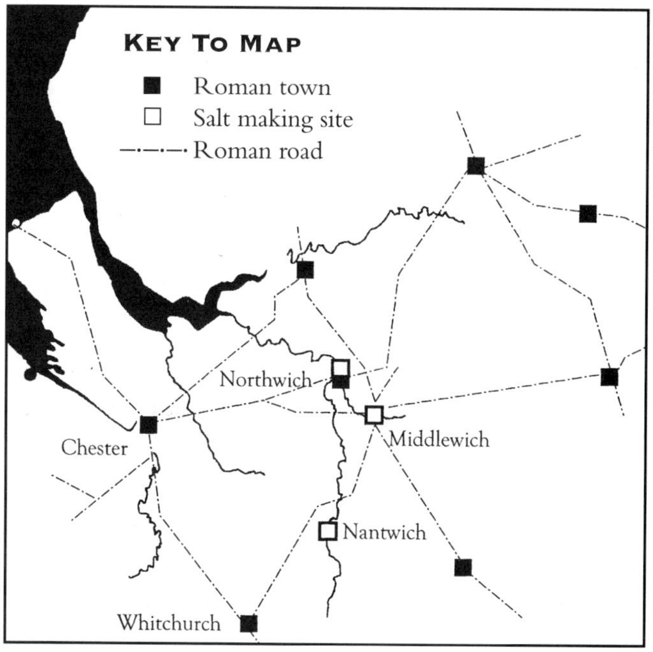

The Roman salt making sites, towns and roads in Cheshire during the 1st Century. Dave Palmer

***Above left:** Re-construction of a Roman saltworks in Middlewich, based on excavations directed by John Bestwick.* ***Above right:** Drawing of a brine pit and salt houses in Saxony by Georgius Agricola De Re Metalica 1556.* Courtesy The Salt Museum, Northwich

brine, with concentrations of salt as much as eight times higher than that of sea water. In earlier times, natural brine springs existed along the edges of the river valleys within the Cheshire Plain and it was from such sources that man first began to make use of the freely available salt.

The earliest confirmed use of these springs for the extraction of salt was between the 1st and 4th centuries A.D., during the Roman occupation of Britain. Remains of workshops dating back to that time have been uncovered in Middlewich, (which the Romans referred to as 'Salinae'), whilst lead pans, which were used to extract the salt from the spring water by evaporation, have been found in Nantwich and Northwich. All three of these 'salt towns' are on low-lying land in river valleys, presumably on the site of earlier natural brine springs. Nantwich and Northwich are on the River Weaver, whilst Middlewich is on the River Dane, a tributary of the Weaver with a confluence in the centre of Northwich (for those with an interest in etymology, the term 'wich' is an old English word that refers to the salt workshops or houses that once existed in the area). The principal uses of the salt at this time would have been for food preservation and flavouring, tanning of leather and dyeing. Transport of salt through Roman-occupied Britain would have been by means of horse-drawn cart using the excellent system of roads set up primarily for the rapid deployment of troops and also by river transport.

The earliest written record of salt extraction in Cheshire is found in the Domesday Book in 1086, from which it would appear that the salt industry in Middlewich, Nantwich and Northwich was already well established. The Domesday Book, however, fails to give any information regarding how the salt was extracted from the brine and it is not until the 16th century that a written description of extraction techniques was available courtesy of William Camden, in his Survey of Britain. The Survey makes reference to the existence of pits at Nantwich and Northwich, from which brine was carried in leather buckets, before being distributed to the salt houses by means of gravity along wooden troughs. At the salt houses the water would have been boiled away by heating the brine in shallow pans over wood or coal fires. Each ton of white salt produced by means of this process required the use of approximately 2 tons of coal or its equivalent in wood.

It is a reasonable assumption that the first of the pits referred to by William Camden would initially have been excavated to assist with the collection of brine from the natural springs. These may subsequently have been deepened in order to follow the groundwater table as the springs dried up. There is little doubt, however, that the plentiful supply of wild brine just below the ground surface would have led to the excavation of other brine pits at new locations in the area not related to springs.

The first reported incidence of brine pumping was not until 1636, when it would appear that manual pumps were employed in Nantwich to raise the wild brine from the pits with distribution to the salt houses, as before, by means of wooden troughs. It was presumably not long after, that the pressures of commerce led to the introduction of mechanically driven pumps to raise the brine, especially as such pumps were already in use in other industries elsewhere in Britain. It was brine pumping, a relatively uncontrolled method of bringing the salt deposits to ground level, that resulted in huge underground caverns. These, not altogether surprisingly, periodically collapsed due to the weight of overlying material, which resulted in sudden and dramatic subsidence at ground level. To take account of this, houses were often built with timber frames so as to permit jacking or raising after settlement. Ground subsidence in rural areas led to the formation of artificial lakes, locally referred to as 'flashes'.

Up until the 17th century, salt was still being transported overland, as in Roman times, either by cart or packhorse, to neighbouring counties or beyond. There was already a growing demand for salt, both from refineries on Merseyside and merchants in Liverpool, for export abroad. By all accounts, however, the road system was then less effective than it had been nearly 1,500 years earlier in Roman times, the roads being in a poor state of repair after many years of neglect.

Rock salt, as opposed to brine, was first discovered in the area in 1670 by John Jackson of Halton, who was prospecting for coal at Marbury. Presumably very much to his surprise, he brought to the surface white rock salt as opposed to the black coal for which he was hoping. Whether or not Jackson realised at the time the importance of his discovery, it is now recognised as a significant turning point in the fortunes of the salt industry. Within just a few years, rock salt was being mined extensively, both for direct use and to strengthen weak brine to enable the production of white salt.

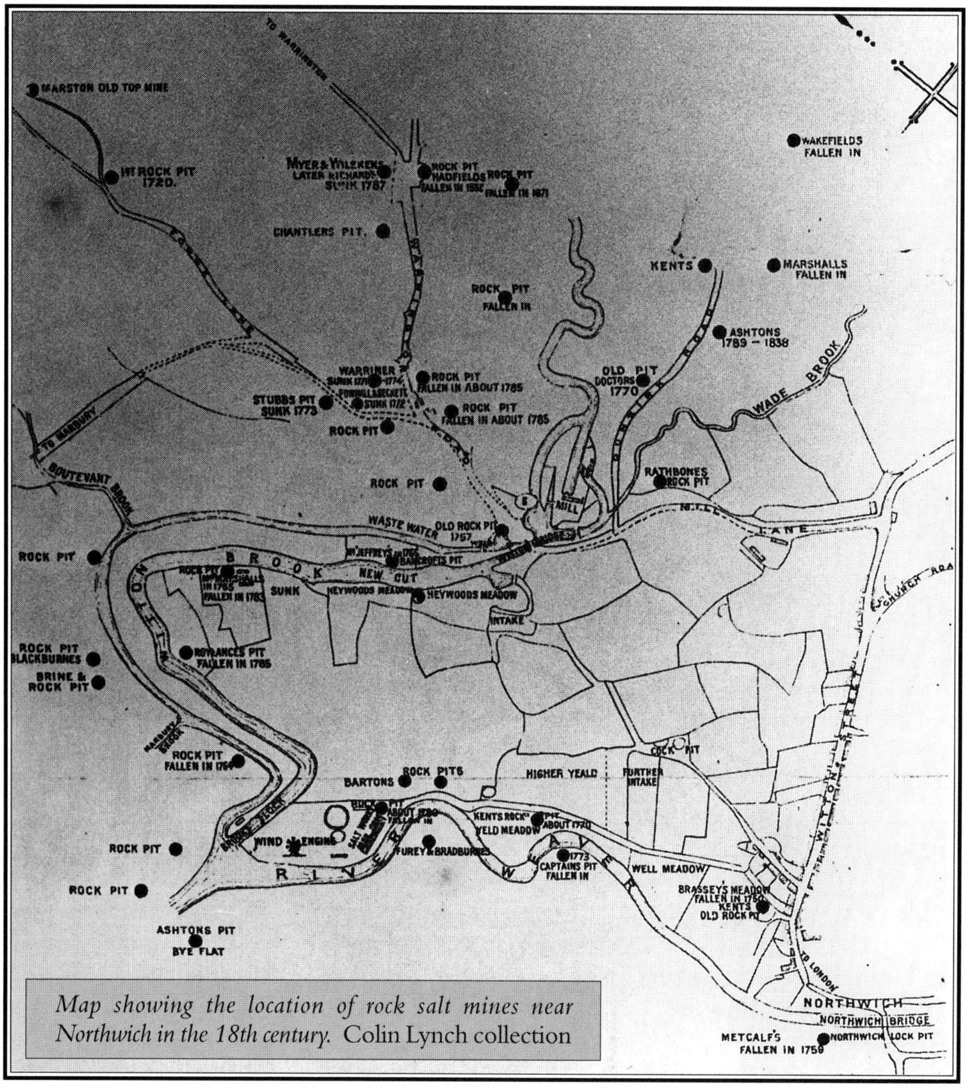

Map showing the location of rock salt mines near Northwich in the 18th century. Colin Lynch collection

Almost undoubtedly, it was the sudden growth in the salt industry, brought about by the plentiful and relatively cheap source provided by the rock salt deposits, which eventually led to the improvement, completed around 1734, of the River Weaver for navigation purposes. Initially, influential Liverpool salt merchants, keen to take full advantage of an expanding market beyond the confines of the immediate area and aware of the poor state of repair of the roads, brought pressure to bear at Westminster for better transport links through the canalisation of the river. After several aborted attempts, an Act of Parliament was finally passed in 1721 which provided the legal basis for the construction of the Weaver Navigation .The preamble to the 1721 Act reads as follows:

'Whereas the making of the river Weaver, alias Weever, alias Wever, navigable for boats, lighters, and other vessels, from Frodsham Bridge, in the County of Chester, to Winsford Bridge, in the same county, will not only be very beneficial and convenient, as well for the carriage of salt and cheese, (the great manufactures and produce of the said county), as of other goods and merchandizes to and from the towns and parts adjacent, but will employ the poor, and be a means to preserve the highways in the said county, and will very much tend to the employing and increase of watermen and seamen, and be for the good of the people.'

The Weaver was effectively made navigable over a distance of some 20 miles, from Frodsham through Northwich to Winsford, thereby providing the salt merchants with much improved transport links between the salt towns and Merseyside. When first constructed, the Navigation comprised twelve timber-built lock structures with several new, shorter cuts and bridges, and providing a minimum draught for vessels of 4 feet 6 inches. The furthest downstream lock at that time was located at the tidal limit at Pickerings near Acton Bridge, with the result that 6 or 7 miles of the river below the lock was not navigable during low tide conditions. This situation, however, was rectified about thirty years later with the construction of a lock at Frodsham.

Prior to the canalisation of the river, it had only been navigable at high tide up to its tidal limit, just 2 or 3 miles upstream of Frodsham Bridge. The salt, which was in great demand on Merseyside and in Liverpool, was having to be transported laboriously by pack-horse and cart from

The British Salt Company's Works at Anderton, with Weaver flats loading salt from chutes in the foreground. View from the River Weaver by J. F. Drinkwater, 1856. Courtesy The Salt Museum, Northwich

the salt towns to Frodsham, before being transferred to boats for shipping up the River Mersey and elsewhere. It should be noted that the maximum burden of a pack-horse was typically 200 lbs., whilst the Weaver Flats, the name given to the sailing vessels which plied their trade on the river, could each carry 35 tons of salt. Is it any wonder, therefore, that the salt merchants were so supportive of the plans for improving the River Weaver?

The 1721 Act was unsuccessfully opposed by landowners on the banks of the river, who saw it as a threat to their livelihoods by the possible disruption of road transport, and infringement of fishing and milling rights. In addition, the white salt proprietors, based in or around the salt towns, viewed the opening of the Weaver Navigation with suspicion. They did not appreciate the potential benefits that the new navigation could bring, not just in relation to the export of salt, but also to the import of much needed coal from Lancashire, for fuelling the white salt extraction process.

During the 18th Century there were two other schemes proposed, both directly related to the Weaver Navigation for which parliamentary approvals were obtained but which, for whatever reasons, were never carried out. The first was to make the River Dane navigable from its confluence with the Weaver in Northwich upstream to Middlewich. This Act was passed at the same time as the first Weaver Act in 1721, but the proposed works, which would have given the Middlewich salt proprietors a direct connection to the Weaver Navigation, were never implemented. The second scheme, for which an Act was obtained in 1738, was for the extension of the Navigation upstream beyond Winsford to Nantwich, but this too was never carried out. The effects of this on the salt proprietors at Middlewich was seemingly not significant, as there was still reasonably good road access to the Navigation at Winsford. On the other hand, the salt proprietors at Nantwich were severely disadvantaged by the failure not to extend the navigation to Winsford and consequently the salt industry in Nantwich did not develop to the same extent.

The early years of the Weaver Navigation were not entirely successful and it was not until the second Weaver Act in 1759 that any significant improvement occurred. This Act, amongst other things, replaced the Undertakers and Commissioners appointed, under the first Act, by Trustees. From this point on, the Weaver Navigation apparently flourished to such an extent a debt of £20,500, which had been accumulated under the first Act, was paid off within twenty years and the first payment of surplus revenue was made to the County soon after. The benefits brought to the Cheshire salt industry by the Navigation is evident from the volumes of goods shipped downstream on a yearly basis after the opening of the Navigation:

It is interesting to note that at the time of the opening of the Navigation, the annual tonnages of both types of salt being shipped downstream were relatively low. Subsequent years up to 1777 saw a steady acceleration in the amount of rock salt being shipped but with a reported levelling off thereafter. Apart from a surge between 1744 and 1747, the white salt shipments grew more slowly than those of rock salt but the growth was maintained beyond 1777, with a tonnage of 100,000 tons a year being shipped down river by the end of the century. The initial fears of the white salt proprietors were clearly unfounded. The discovery of the rock salt deposits, combined with the opening of the Weaver Navigation, had evidently attracted major investments into the region and encouraged significant advances in the white salt extraction process.

GROWTH OF SALT TRAFFIC ON THE RIVER WEAVER IN THE 18TH CENTURY

YEAR	ROCK SALT	WHITE SALT
1732	9,322 tons	5,031 tons
1744	11,744 tons	8,279 tons
1747	13,310 tons	16,101 tons
1764	30,141 tons	18,637 tons
1777	54,176 tons	31,000 tons

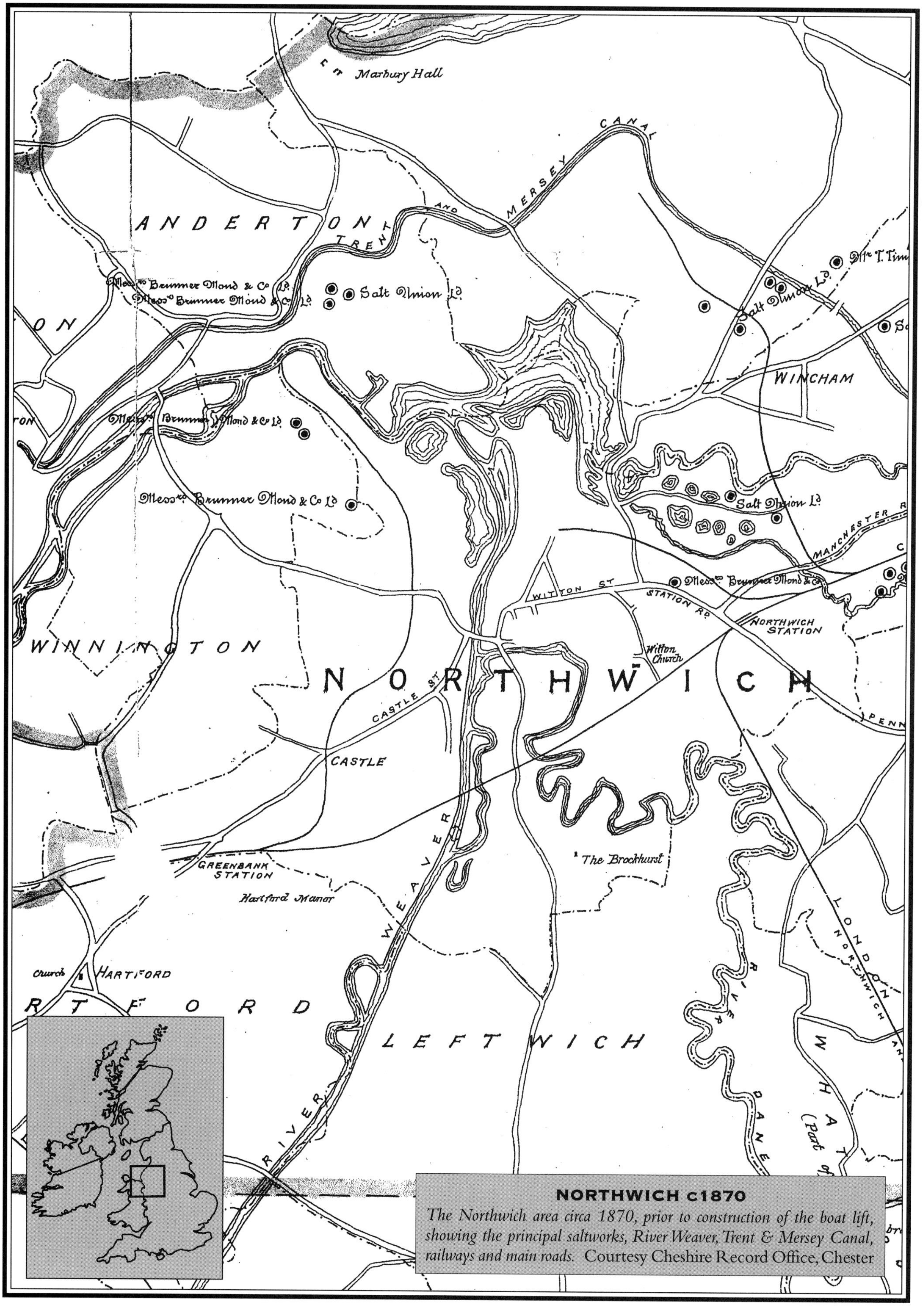

NORTHWICH c1870

The Northwich area circa 1870, prior to construction of the boat lift, showing the principal saltworks, River Weaver, Trent & Mersey Canal, railways and main roads. Courtesy Cheshire Record Office, Chester

The Salt Industry
Mining and Processing the Salt

Top right: *A brine pit, showing the head gear and pump.* **Top left:** *A posed photograph of miners in the Adelaide rock salt mine at Northwich in 1920.* **Middle left:** *The tubs of rock salt were taken to the shaft for hauling out of the mine on horse drawn trams. Note the names and initials scratched on the rock face behind.* **Middle right:** *The water was pumped into salt pans, sited on top of a bank of furnaces. The fires were attended by 'lumpmen', whose job it was to keep the brine on the boil. It took about two tons of coal to produce three tons of salt.* **Bottom left:** *The fine salt produced by this method was broken up in milling machines and bagged for sale. The men on the left are filling bags whilst the women sew them up.* **Bottom right:** *Lumps of salt produced with the aid of moulds.* All courtesy The Salt Museum, Northwich

Top left: The production of coarse or broad salt was a slower process, taking from three to four days and being carried out in much bigger pans. The salt was raked to the edge of the pans by the 'wallers', a hazardous job as the act of throwing the long rakes to the centre of the pan had to be accomplished without overbalancing and falling in. The saltworker pictured here has devised his own novel solution to this problem, with the id of a couple of wooden buckets. The salt was then scooped out and left to drain on wooden hurdles on the floor, before being shovelled into handcarts, as seen, and either taken to the storehouse or tipped straight into waiting river vessels. Mike Day collection

Middle left: Squares of fine salt being stacked in the storehouse. When dry, it was either milled and bagged as loose salt, or was sold directly as squares. Note the tram rails and tram truck on the left, which probably led to the milling machines and bagging room.

Middle right: A view inside a salt storehouses, filled with blocks of moulded salt. Broad salt was stored loose, in another shed.

Both courtesy The Salt Museum, Northwich

Wincham Hall or the Lion Saltworks, on the Trent & Mersey Canal at Marston near Northwich, circa 1905. The works survives today as a museum, complete with the last remaining set of coal-fired salt evaporation pans in the country. Courtesy The Salt Museum, Northwich

THE SALT INDUSTRY

LOADING AND SHIPPING THE SALT

Top left: *The distinctive boats of the local river were the Weaver packets, steam driven vessels constructed of wood or iron and a close relative in design to the flats of the River Mersey. Here* Vale Royal, *a packet built in 1873 as a flat but converted to steam ten years after, is seen loading loose salt at Anderton, circa 1914. The tubs at the rear of the vessel were for loading rock salt and had a 2 ton capacity. Like many packets,* Vale Royal *enjoyed a long life, surviving until 1953 when she was broken up.*
Photo H. Davis

Top right: *Bagged salt being winched aboard a packet at Anderton in the 1950s. The holds of the packets had to be lined and watertight, as loose salt is extremely corrosive and bagged salt had to be kept dry.* Cyril Bailey collection

Middle left: *White salt being unloaded by hand from a narrowboat into carts on the Trent & Mersey Canal at Anderton. This would then be tipped down the chutes into waiting packets on the River Weaver below.* Courtesy The Salt Museum, Northwich

Bottom left: *A posed view at the Lion Saltworks of Henry Ingram Thompson, at Marston, circa 1910. Broad or coarse salt has been loaded aboard the company's own narrowboat for transport along the Trent & Mersey, probably to the chutes at Anderton; another vessel awaits loading behind.* Courtesy The Salt Museum, Northwich

White salt being loaded by chute from narrowboats on the Trent & Mersey Canal above and spread by hand into a packet in the Anderton Basin. As can be deduced from these photographs, the salt industry was very labour intensive and remained so right up until the middle of the 20th century.
Courtesy The Salt Museum, Northwich

The Weaver packet Burmah, *built as a sailing flat in 1875 and converted to steam in 1882, seen at Winnington circa 1935. This posed photograph is not quite what it seems; loaders are seen with barrows of both bagged and loose salt but the packet would never load a mixed cargo like this. Another cart can be seen on the tipping stage, although the vessel's hatch appears to be closed! Note also the characteristic outside framing of the salt sheds, which left the inside walls flush so loose salt did not gather in nooks and corners. Another packet can be seen beyond the* Burmah *and the Salt Union's launch* Pacific *appears on the right.* Courtesy Ellesmere Port Boat Museum

The Salt Industry
The After Effects: Subsidence

SHOPLIFTING IN NORTHWICH. IInd STAGE
Photo Jeffries Evans Publish.

The extraction of so much brine and rock salt over the centuries, coupled with the low-lying nature of much of the area and the growth of housing, shops and so on for the people who worked in the industry, brought catastrophic results at the end of the 19th and in the early years of the 20th centuries, as these dramatic pictures show. Subsidence began to effect Northwich and the surrounding area, bringing scenes of unbelievable devastion, and a large number of commercial postcard views were published depicting these events.

Top left: *The caption on this postcard says it all; this house has sunk about 5 feet.*
Courtesy The Salt Museum, Northwich
Top right: *Northwich High Street had to be raised, shops and all, several feet every few years to keep it above river level. The subsidence caused major disruption in the street, with damaged mains also having to be regularly repaired, as seen here, but life – and business – had to go on.* Mike Day collection
Middle left: *A row of terraced houses in London Road, Leftwich. The comic appearance belies the heartache caused to poor working families, many of whom whould still be putting in long shifts in the industry which had caused the devastation.* Colin Lynch collection
Bottom left: *Shops and houses were jacked up like this to counter the subsidence. The ground underneath was then packed with earth and stones before the street itself was raised. Within a couple oy years or so, they would have to do it all again. Looking at these pictures, the fortitude, ingenuity and tenacity of the local population can only be admired.*
Courtesy The Salt Museum, Northwich

Top: *A rather severe result of ground subsidence in Northwich in 1894, again due to the uncontrolled brine pumping in the area. The building, Castle Chambers, in use as offices, leans drunkenly at about 30⁰ off the perpendicular. The watching crowd and the removals men have all been distracted momentarily by the presence of the photographer recording the scene for posterity.* Courtesy The Salt Museum, Northwich

Centre: *A 'flash', or waterfilled hollow, adjacent to the Witton Brook, near Northwich, circa 1910, caused by ground subsidence due to brine pumping. This one, locally known as Ashton's Flash after an old rock salt mine which had closed in the late 1830s, also caused Dunkirk Road, adjacent to the fence running across the picture, to be submerged. A Weaver flat is in the left foreground and one of the Salt Union's saltworks appears in the background.* Colin Lynch collection

Below: *A breach in the Trent & Mersey Canal near Northwich in July 1907, as a result of ground subsidence. A crowd has gathered to watch the efforts of those trying to rescue the stranded boats. From 1845 until the 1923 Railway Grouping, the canal was in the ownership of the North Staffordshire Railway and it is one of their Engineers Department narrowboats which features in this photograph, stuck in the drained canal.* Mike Day collection

THE POTTERY INDUSTRY
PACKING & TRANSPORTING THE GOODS

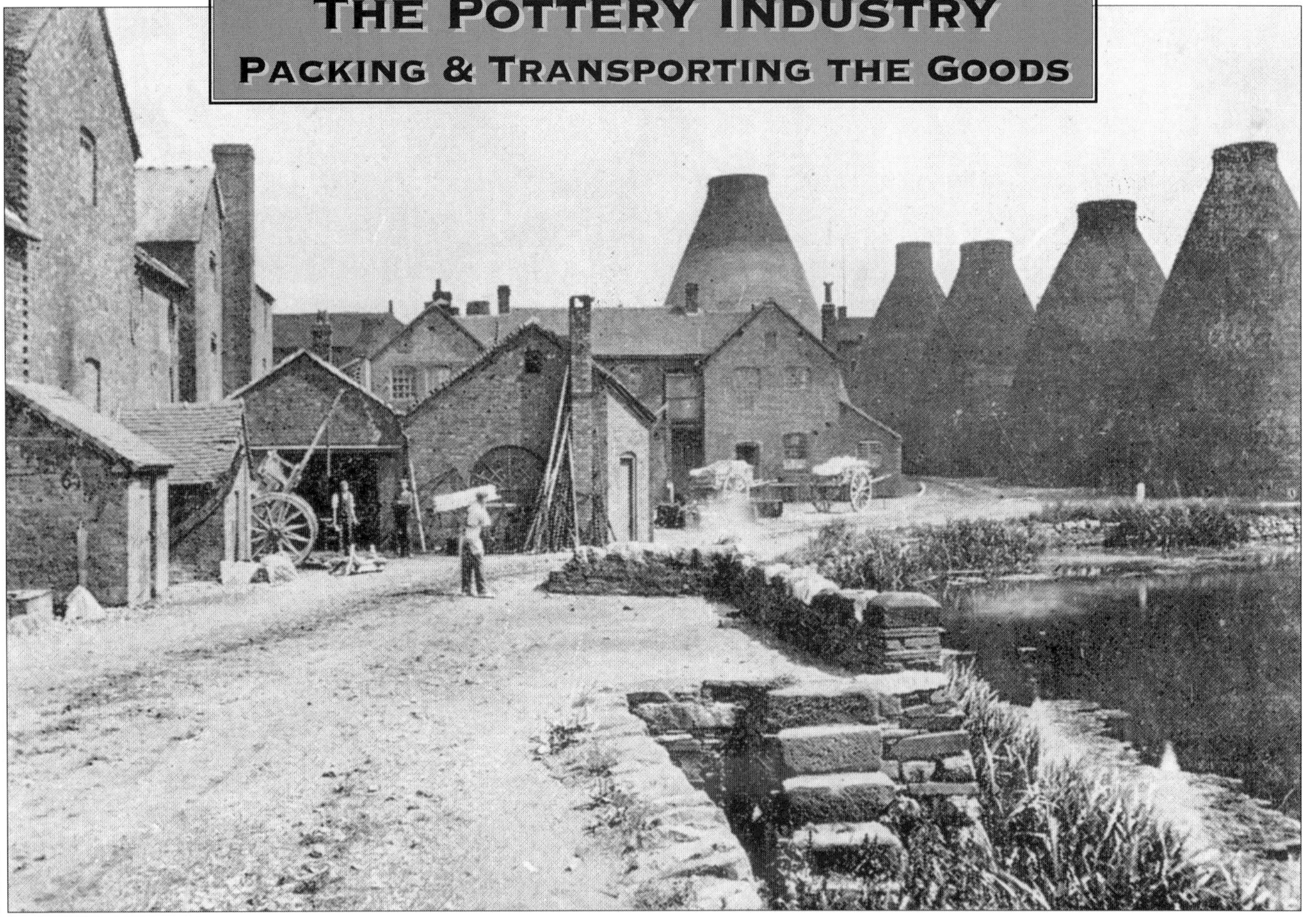

The Wedgewood factory alongside the Trent & Mersey Canal at Etruria, in the heart of the Potteries. Josiah Wedgewood was the driving force behind the building of the canal, thus linking his pottery with markets both nationally and beyond these shores, as well as facilitating the inward transport of raw materials. Courtesy The Potteries Museum and Art Gallery, Stoke-on-Trent

A crate packer at work in the Potteries. This was the traditional way of transporting finished china-ware on the canal and river. Courtesy The Potteries Museum and Art Gallery, Stoke-on-Trent

Pottery being packed with straw into barrels around 1905. Canals were the ideal transport medium for chinaware and pottery, where careful handling and a smooth journey were generally more important than speed of delivery.
Courtesy The Potteries Museum and Art Gallery, Stoke-on-Trent

A laden horse-drawn narrowboat wends its way along the Trent & Mersey Canal, with the Wedgwood factory and offices forming the backdrop, in the early years of the 20th century. The horse is climbing a raised section of towpath, probably to cross over a branch arm leading off the main canal.
Courtesy The Potteries Museum and Art Gallery, Stoke-on-Trent

A later view, probably circa 1950s, showing a motor narrowboat negotiating its way along a somewhat neglected-looking section of the Trent & Mersey Canal in the Potteries.
Courtesy The Potteries Museum and Art Gallery, Stoke-on-Trent

A Mersey Weaver Company narrowboat, loaded with either flints or coal (it is alongside the boilerhouse), at a warehouse on the Trent & Mersey Canal in the Potteries. A sister craft is moored in front. Courtesy The Potteries Museum and Art Gallery, Stoke-on-Trent

Clay and the Trent & Mersey

Pottery making in North Staffordshire is said to have begun in Elizabethan times and relied principally upon the Keuper Marl beds. Like the salt deposits in Cheshire, Keuper Marl had its origins in the Triassic Period of geological history, around 200 million years ago. During this period, as the theory goes, huge inland lakes existed in the area and these resulted in the deposition of silts and clays. Over the ages, the silt and clay beds at the bottom of the lakes increased in thickness and developed, by the process of compaction, into what is referred to today as Keuper Marl. In places the Keuper Marl stratum is up to 2,500 ft. thick.

In the Elizabethan era, the late 16th century, pottery making would have been undertaken on a local, small-scale basis in those parts of the country which had a ready source of suitable clay and adequate fuel for firing purposes. Initially trees, which once covered most of the country, would have been the principal source of fuel for the kilns. However, as the demand for pottery-ware increased and the industry became more specialised, pottery making gradually became concentrated into those parts of the country that had particular commercial advantages. So for example, the reason for the expansion of the pottery industries at London, and other urban areas such as Bristol and Norwich, would have been the ready access to the marketplace, both at home and overseas. In the case of the Staffordshire Potteries, the reason for success was probably the locally available coal deposits. These are thought to have first been used in the 17th century, as a fuel to replace the fast diminishing stocks of timber. So it was that, by 1710, the town of Burslem had become a prominent pottery centre, probably the largest in Britain, and had acquired a reputation for both skill and craftsmanship.

In addition to Burslem, however, there were five other towns huddled together in the Trent valley, which collectively became known as The Potteries, in which pottery making was, and still is, a principal activity. From north to south the six towns are Tunstall, Burslem, Hanley, Stoke, Fenton and Longton. With the success of the pottery industry over the years, these once separate towns gradually expanded and merged into a larger urbanised area which today is referred to as Stoke-on-Trent. In those early times fierce competition existed between the towns, each tending to develop its own speciality in pottery-ware.

The main disadvantage, however, which faced the Staffordshire potters in the 18th century, was the poor transport links to the rest of the country and to the coastal ports for access to the overseas market. At that time, the Trent was navigable as far as Nottingham but several attempts in the early 1700s to extend the navigation further upstream were successfully opposed by landowners. In time an Act was passed enabling the river to be made navigable, albeit unreliably so, as far upstream as Burton. By that time, the Weaver was navigable between Frodsham and Winsford.

By the middle of the 18th century, pottery making in North Staffordshire had developed from cottage to factory scale and, by so doing, had become more specialised. The local stocks of raw materials were found to be either insufficient or unsuitable. The potters in the area had turned to other parts of the country to source their raw materials.

By the 1740s, china clay from Cornwall was being shipped up the west coast, then via the River Mersey and the River Weaver, before being carried overland to the Potteries by waggon and pack-horse. In a similar vein, flintstones from Kent were being shipped up the east coast either to ports such as Hull, or up the River Trent to Gainsborough and other towns. The last leg of the journey, however, was again overland by waggon or pack-horse. The reverse procedures were used for exporting the finished pottery-ware to outside markets. The cost of the transport, particularly the overland sections, was a huge burden on the Staffordshire potters and, as a result, pressure began to grow for a canal linking the Potteries to the river navigations.

The initial move for a canal came from Mr. Hardman, a Liverpool merchant, who organised a survey of a route in 1755 but this apparently came to nothing. Not long after, however, two very prominent figures came onto the scene. In 1765, Josiah Wedgewood, a leading pottery manufacturer in the area, started his own campaign for a canal to link the Potteries with the River Trent to the south east, and the Mersey via the Bridgewater Canal in the north west. As a consequence, James Brindley was appointed as Engineer to the project and proceeded to survey a route for what he referred to as the 'Grand Trunk' but which subsequently came to be known as the Trent & Mersey Canal. Brindley had previously earned his reputation as a canal pioneer working for the Duke of Bridgewater, on the canal system for his collieries in North Cheshire.

Despite vehement opposition from the river navigations and overland traders, an Act of Parliament was eventually passed on 14 May 1766 enabling the construction of the canal. The proposed route was from Rugeley, on the River Trent, north westwards through Stone, Trentham, Stoke, Middlewich and Northwich, to join the Bridgewater Canal at Preston Brook. From there it was just a short cruise to the River Mersey at Runcorn. Coincidentally, on the same day an Act was passed for the Staffordshire & Worcestershire Canal, which would link the Trent & Mersey Canal at Great Heywood with Wolverhampton and eventually the River Severn at Stourport.

Work on the construction of the Trent & Mersey, under Brindley's control, was commenced within a few years of the passing of the Act and the section south of Stoke was opened to traffic in October 1772. The next section, between Middlewich and Stoke, which includes the impressive Harecastle Tunnel near Kidsgrove, was completed by September 1775. The final section, however, between Middlewich and Preston Brook, which involved the construction of tunnels at Barnton, Saltersford and Preston Brook, was only completed in May 1777. The total length of the canal was 93 miles. Perhaps not surprisingly, on the opening of the canal along its full length, it was described as the greatest civil engineering works so far built in England.

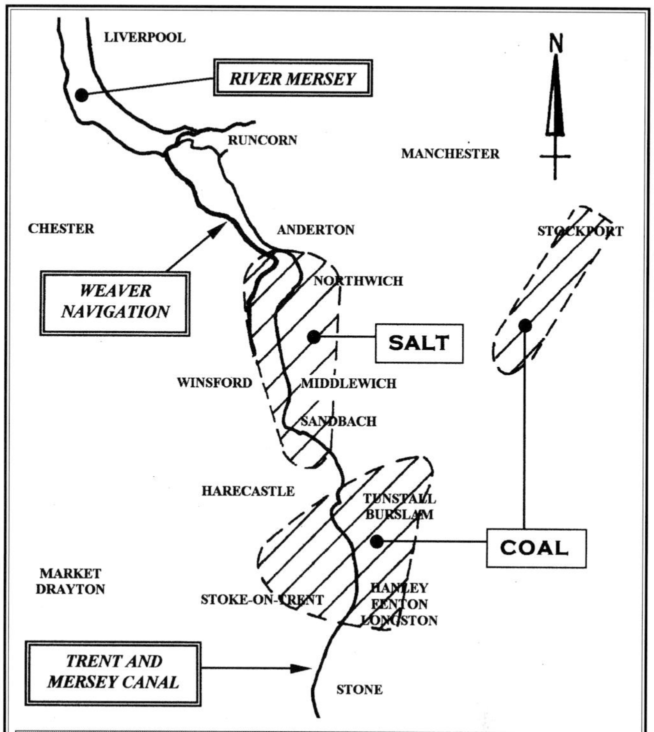

Plan of the Salt Towns and the Potteries, showing the juxtaposition of the Weaver Navigation and the Trent & Mersey Canal with the salt and coal deposits. Sue Lamb/Dave Palmer

The Potteries benefited enormously from the new canal, as did other industries in the area, such as brewing, salt making and coal mining. The much reduced transport costs brought greater prosperity to North Staffordshire and resulted in a major population growth in the region.

And so the scene was set, with the Weaver Navigation going from strength to strength with the rapid expansion of the Cheshire salt industry, and the Trent & Mersey Canal also thriving from the carrying of goods to and from the Staffordshire Potteries, for the development of the Anderton Basin. It would not be long before the Weaver Trustees took the decision to build a connection between the two waterways.

INFORMATION SOURCES FOR CHAPTER 1

The Salt Museum, Northwich - Information Pack, *including pamphlets on* Roman Salt Making, Salt Making in Domesday Cheshire, The Geology of Salt *and* Transport in the Salt Industry. *Salt Museum Publications, Cheshire Libraries and Museums.*

Canals of the West Midlands, *by Charles Hadfield, published by David & Charles.*

Notes on the River Weaver Trust, *two papers read by Thos. Ward, J.P. before the Northwich Debating Society in 1887.*

The Weaver Navigation, *by Harry Wardale.*

The Potteries, *Shire Album No. 62, published by Shire Publications Ltd.*

The Trent & Mersey Canal, *by Peter Lead as part of a series of publications entitled Historic Waterways Scenes.*

Trent & Mersey Canal Cruising Guide, *published by Waterways World.*

Miners, Quarrymen and Saltworkers. *Edited by Raphael Samuel. Routledge & Kegan Paul, 1977.*

A 1930s photograph showing cargo being transshipped in the Anderton Basin. Here, bagged salt, brought down from the Trent & Mersey, via the boat lift, is being transferred to one of Henry Seddon's steam packets (possibly the Weaver Belle*) by means of the latter's derrick. Note how the stern of the narrowboat lifts out of the water as unloading progresses. Prior to construction of the lift, transshipment of cargoes at the basin was effected by means of chutes and inclines.* Cyril Bailey collection

CHAPTER 2
THE ANDERTON BASIN

'On 7th April 1791, the Committee [of the Trustees] *decided to acquire land at Anderton adjoining the canal from Sir John Stanley and his son. In the following month they agreed to negotiate with Lord Mansfield for about a statute acre of land in the same place.'*

Extract from *The Navigation of the River Weaver* by Willan (1952)

With the Weaver Navigation going from strength to strength as a result of the rapid expansion of the Cheshire salt industry, and the Trent & Mersey Canal thriving principally as a result of carrying goods to and from the Staffordshire Potteries, it was only a matter of time before a link was developed between the two waterways. Accordingly, as a direct response to pressure from local businessmen, the decision was taken by the Weaver Trustees at the end of the 18th century to create some form of connection between the canal and the river.

* * * * * * * * *

Prior to the construction of the Trent & Mersey Canal, the Weaver Navigation profited significantly from the trade both to and from the Cheshire salt fields and the Staffordshire Potteries. Not surprisingly, therefore, when the Trent & Mersey Canal was first mooted in the mid 1700s, the Weaver Trustees took fright at the obvious threat to their income and put up vigorous opposition to the canal. One of their opening gambits against the proposed canal was, with the support of '*the leading men in Cheshire*', to put forward plans for construction of their own canal, linking the River Weaver with the River Trent, between Nantwich and Wilden Ferry, passing through Eccleshall and Stafford. This, however, did not gain much support from the Staffordshire potters, presumably as it would have been too far to the south of the Potteries district. The Trustees then put forward the idea of a canal between Northwich and Manchester, via Macclesfield and Stockport. The intention of this scheme was probably to prevent the proposed Trent & Mersey Canal from linking up with the Bridgewater Canal, thus stopping canal traffic from reaching the Mersey without the easier option of joining the Weaver Navigation.

Conflict between the Weaver Trust and the promoters of the Trent & Mersey Canal continued for a number of years, until the matter was seemingly settled in 1766 by the passing of an Act of Parliament in favour of the Canal and the rejection of the Trust's alternative. However, even with the new Act on the statute book, there remained considerable hostility between the Trustees and the canal proprietors. This was not altogether surprising, as the intended route of the canal was diagonally across the country, creating a link between the Trent near Shardlow and the Mersey at Runcorn, via the Bridgewater Canal. As such, it would pass through the industrialised areas of Stone, Trentham, Stoke, Middlewich and Northwich. The main difficulty, as far as the Trustees were concerned, was that the proposed canal would come between the Weaver Navigation and the countryside to the east from which it had hitherto received trade, including the whole of the Potteries, the salt fields of Middlewich, and those to the north of Northwich in the Marston area.

When the Trent & Mersey Canal was eventually completed in 1777, the river and the new canal faced each other over the fields of Cheshire rather like the front line trenches in northern France during the First World War. By this time, the primary objective of the Weaver Trustees had become that of attracting the remaining trade from this 'no-man's land' between the two waterways, westwards onto the river.

As had been feared by the Trustees, the opening of the canal did result in a reduction in traffic on the Navigation but, by all accounts, it was not as serious as originally envisaged. The main loss was associated with raw materials heading for the Potteries, such as clay and flintstones, and the finished products being carried away from there. On the positive side, the amount of coal carried on the river to the salt districts, as well as salt transported away from Northwich, were largely unaffected. Nonetheless, the Trustees were still eager to find ways of replacing the lost trade. Consequently, as well as continuing the process of improving the Navigation itself, they began to invest in areas of business that were, hitherto, new to them. These included the construction and improvement of roads leading to the river – for example, between Winsford and Middlewich, and between Wincham and the Witton Brook – and the construction of wharves at Witton Brook.

About the same time the Trustees began a more aggressive campaign to tempt carriers to transfer their cargoes from the canal to the river for the journey out to the Mersey. In

1788, largely as a result of pressure from salt proprietors at Middlewich but perhaps also with encouragement from the pottery manufacturers, the Trustees instructed their Engineer, Mr. John Johnson, to investigate ways of creating a positive link between the two waterways. In March 1789, the Trustees formally approved Johnson's initial proposals:

'for making quays and engines with communications and other conveniencys'

and gave instructions for negotiations to start for the purchase of suitable land. It is not clear whether any particular site had at that stage been chosen, but the most obvious locations for a connection were to be found at Anderton, Barnton or Saltersford, where the two waterways came within just a few hundred yards of each other. Between Northwich and Dutton, a distance of nominally 5 miles, the route of the Trent & Mersey Canal is very close to the River Weaver. The two waterways run roughly in parallel in an east/west direction, separated vertically by a height of just over 50 feet.

A year or so later, two more specific proposals were put to the Trustees by Johnson. One included what was rather obscurely described as '*cutting a sluice across the towing-path of the canal*' whilst the other was more simply described as for '*wheeling goods over the towing-path*'. Not surprisingly, given the ongoing hostility, the canal company refused to give consent for the water-based connection and consequently the Trustees elected to proceed with the latter option, for which, apparently, consent from the canal company was not needed. Subsequently, inquiries were made with wharf owners elsewhere in order to ascertain the best means of transshipping goods overland between waterways.

Soon after, the decision was taken to locate the overland connection at Anderton, and negotiations were commenced with relevant land owners. However, progress was slow, principally due to hindrance from the canal company, and it took a further 18 months before the Trustees took possession of land from Sir John Stanley. It was on 6 June 1793 that the Trustees were informed:

'that the canal company did not mean to defend the ejectment brought to recover the possession of the piece of land in Anderton purchased by the Weaver Trustees from Sir John Stanley, Bt., but would give up the possession thereof immediately, reserving to the canal company such powers as they are intitled under and by virtue of the Act of Parliament for making the said canal.'

Having finally gained possession of the land at Anderton, Johnson was almost immediately instructed to start work on the connection between the waterways by:

'pulling and breaking down so much of the hedge or fence . . . which divides the said towing-path from the said piece of land'

The principal component of the works that were carried out in 1793 under the direction of Johnson, was the excavation of a basin, leading from the north bank of the river to the base of the escarpmen, to within 50 yards of the canal. The vertical sides of the basin were formed by the use of local sandstone blockwork.

It was around this time that Johnson made a visit to Coalbrookdale in Shropshire with the intention of:

'examining and informing himself fully of the nature and method of making the different railed roads and other works there for shipping goods into vessels on the river Severn.'

It is thought that he would have inspected the inclines and related structures at Coalport, between the Shropshire Canal and the River Severn. These were used for the transfer of coal into boats on the river below, making use of wagons on the inclined planes.

Over the following years, several further developments took place at the Anderton Basin, including the erection of timber storage buildings and wharves, timber wheeling stages and salt chutes, and a small house for the Trustees' manager. Until then, it appears most of the transshipment was downwards, from the canal to the river, principally salt from the Middlewich area but also pottery-ware. In 1796, however, two cranes were erected, in the hope that these would tempt carriers to bring the raw materials destined for the Potteries along the Weaver before being lifted up to the canal. By all accounts, though, it was unlikely that the upwards movement of goods was of any great quantity compared to those moving in the other direction.

The total cost of the works carried out by the Trustees in the first five years of the Anderton Basin (*i.e.* April 1793 to April 1798) was just £2,077 12s – less than the cost of a new lock. During a similar 5 year period, (*i.e.* April 1795 to April 1800), records indicate that a total of 17,674 tons of white salt from Middlewich had been transshipped from the canal to the river at Anderton. However, it is not recorded how much income this traffic generated for the Trustees.

By the turn of the century, two salt chutes and an incline, probably similar to the Hay Incline at Coalport, had been built on the steep bank separating the canal and the basin. Rock salt was then being manually offloaded from barges moored in the canal and barrowed across the towpath, onto wooden platforms, or gantries, projecting out across the river valley. From the platforms, the salt was tipped down wooden channels directly into river boats waiting in the basin below. The incline almost certainly comprised two parallel rail tracks up the steep bank. Wagons would have been raised and lowered in unison, by means of the upper truck being loaded more heavily than the lower; thus the lighter one was drawn up. The incline enabled goods other than loose salt to be transshipped.

Following a deputation '*from gentlemen concerned in the pottery trade*' wishing to see an expansion of the movement of china clay and pottery-ware between the two waterways in the early years of the 19th century, the Trustees authorised further development of the facilities for transshipment at Anderton. More land was purchased from Sir John Stanley and in 1801 a new quay was built for handling clay, with the subsequent erection of cranes and stages. The pottery traders were consulted on the proposals prior to the works being carried out. Records indicate

The Wombridge or Trench Incline on the tub-boat section of the Shrewsbury Canal may have been one of those visited by Johnson on behalf of the Weaver Trustees around the end of the 18th century. The incline was 670 feet long with a vertical rise of 75 feet. Similar, although much smaller inclines were built at Anderton to facilitate the transshipment of goods between the two waterways. This late 19th century photograph shows the engine house and basin at the top of the incline. A full tub boat can be seen moored by the wharf, whilst another seemingly empty one has just completed the ascent.

The view back up the incline from the bottom basin, with an empty tub about to ascend. Note how basic they were in design and their relatively small size. Johnson would probably have recognised that this system would have been unsuitable for larger vessels as used on the Trent & Mersey Canal. Both courtesy The National Waterways Museum, Gloucester

Two photographs of the Foxton Inclined Plane, on the Grand Union Canal near Market Harborough in Leicestershire, one of the 'Seven Wonders of the Waterways', as is the Anderton Boat Lift. This view is thought to be at its formal opening on 10 July 1900. Two immaculately prepared narrowboats, with crews also suitably 'spruced up' for the occasion are posed entering one of the caissons. Although built long after the Anderton Lift, Johnson would have been aware of this type of inclined plane as others existed elsewhere, most notably at Blackhill, on the Monkland Canal, in Scotland.

The two caissons in operation, both laden with a pair of narrowboats, also on the opening day. The massive size of the operation can be appreciated and the space it required would probably have made a similar structure inappropriate for the more confined site at Anderton. The main problems encountered in use were with the rails, which were too flimsy and had to be replaced with larger section, and hydraulic cylinders fracturing in frosty conditions. Foxton represented the zenith of steam powered incline construction in Britain and was successful in operation, taking 12 minutes to pass boats instead of the 80 minutes taken on the flight of locks it by-passed. Unfortunately, shortly after it was built, the Board of the Grand Union Canal decided to rebuild their flight of locks at Watford as narrow locks, thus precluding wide boats from using the busiest section of their waterway. This negated Foxton's advantage of being able to carry wide boats. With anticipated traffic from the north failing to materialise, the incline quickly became something of a 'white elephant', closing after a very short life in November 1910; it worked occasionally after that, the machinery not being scrapped until 1928. Both courtesy The National Waterways Museum, Gloucester

that, between 1801 and 1808, the Trustees expended a further amount of £3,417 on development works. Clearly, the income resulting from the transshipment of goods at the Anderton Basin, whether direct or indirect, was of great significance to the Trustees.

★ ★ ★ ★ ★ ★ ★ ★ ★ ★

It might now be helpful to take a wider view of the situation. By 1800 the Weaver had been navigable for nigh on eighty years, for most of that time under the jurisdiction of the Trustees who managed the Navigation *'for the public good'*. The distance from Northwich to the Mersey Estuary, initially the main export route for Cheshire salt, was nominally 14 miles. At that point in time, the tidal limit of the river was at Frodsham Bridge, with the consequence that boats could not gain access to the Mersey during all states of the tide. However, when tidal conditions at Frodsham permitted, the river boats were capable of proceeding to the port of Liverpool, thereby avoiding the expensive operation of transshipment of goods.

In comparison, by 1800 the Trent & Mersey Canal had been open for just over 20 years and was thriving as a result of the growth of the Potteries in Staffordshire. The canal gave access to barges to the Mersey at Runcorn, via the Bridgewater Canal, a distance of some 15 miles from Northwich. At Runcorn, however, the goods had to be transshipped to vessels more suited to conditions in the estuary. Furthermore, the Duke of Bridgewater operated a dock, at a fee, for boats moving between the estuary and the canal.

The overall cost of carrying goods between Anderton and Liverpool by means of the Weaver Navigation, was competitive with that of using the alternative route on the canals via Runcorn, especially if the Trustees minimised the charge to carriers for using the Anderton Basin. By such means, the Trustees managed to attract a significant volume of both salt and pottery-ware off the Trent & Mersey Canal and onto the Weaver.

In the light of the above circumstances, to increase further the attractiveness of the Weaver, the Trustees promoted a Bill through Parliament for the construction of the Weston Canal between Frodsham Bridge and deep water moorings at Weston Point, which would enable access to and from the river at all states of the tide. Against opposition from the canal company, the Bill was enacted in 1807 but they did manage to have inserted additional clauses which expressly forbade the transshipment of pottery materials and goods at Anderton. This embargo remained in place, at the obvious detriment of the Weaver Trustees for 18 years or so. In time, it was repealed by another Act of Parliament in 1825, the preamble of which read as follows:

'And whereas the prohibition and restrictions hereinbefore recited have been found highly prejudicial to the public, by impeding and rendering more expensive the conveyance of goods, wares, and

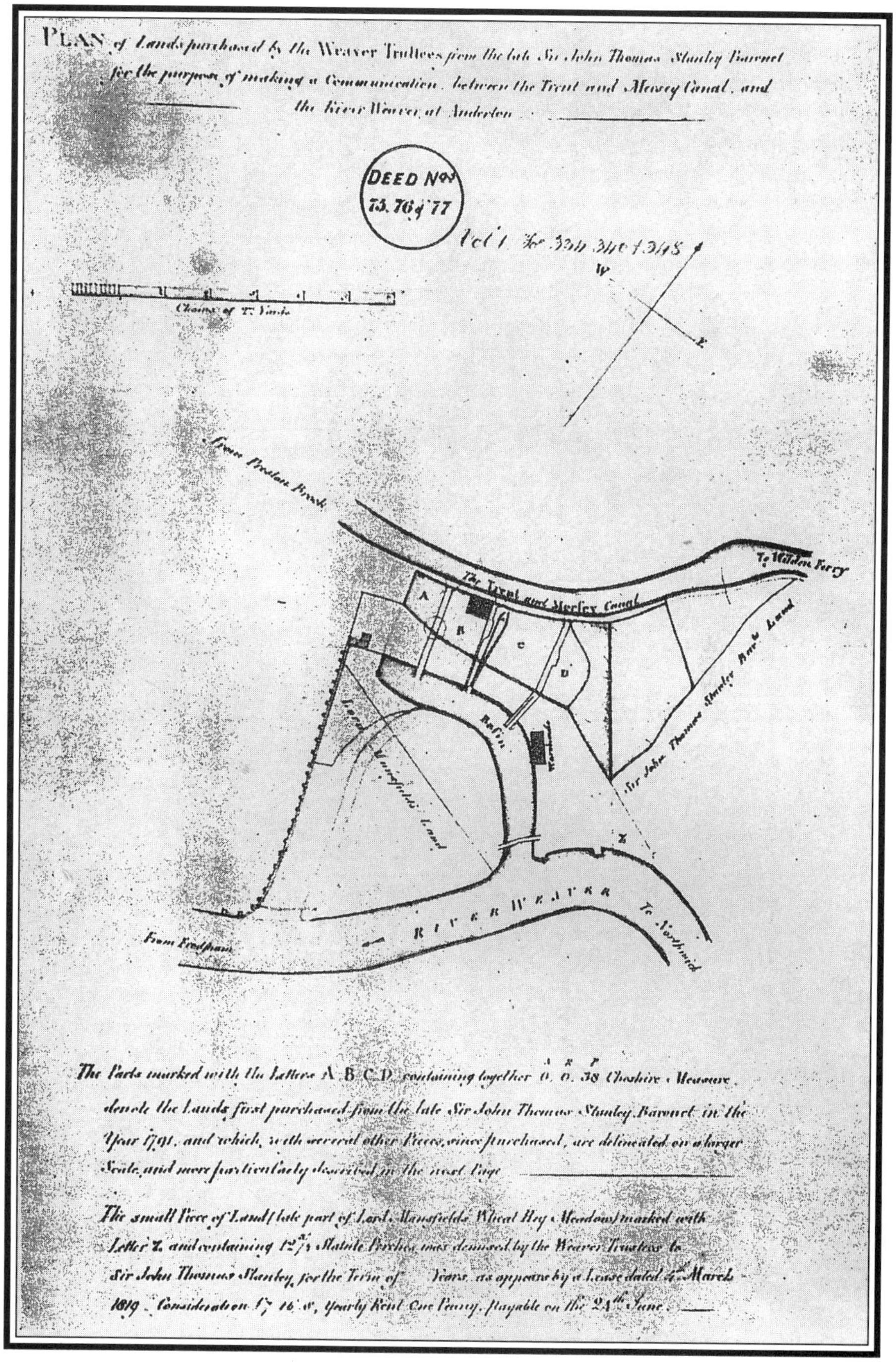

An extract from the deed showing the land purchased at Anderton by the Weaver Trustees from Sir John Thomas Stanley in 1793. There appears to be three connections, presumably two salt chutes and a tramroad, and one warehouse already in use.

merchandise from various parts of the country, near to or in the vicinity of the River Weaver Navigation and the Trent and Mersey Canal, and the communication for trade and merchandise between those parts and the town and port of Liverpool . . . It is expedient that the said restrictions and prohibitions should be repealed . . .'

With a cessation of the main hostilities between the Trustees and the canal company, marked by the 1825 Act, the transshipment of clay, flint and finished pottery-ware at Anderton was recommenced and rapidly grew in volume over the following years. In response to the increased demand, in 1831 the Trustees authorised the construction of a second entrance to the basin from the river, about 150 yards downstream of the original entrance, with the consequence that a loop with a central island was created. By this time, the facilities for transshipment included three salt chutes, two inclined planes and warehousing on the eastern edge of the basin. The most recent of the inclined planes probably consisted of a single track and was operated by steam engine. It was located at the north west corner of the basin.

By the middle of the 19th century, the Anderton Basin had become a major focus for canal and river traffic in the north west of England, with an annual tonnage of over 30,000 tons of pottery goods alone (either raw materials or finished goods) being transshipped. Several commercial carriers had established bases at the basin to take full commercial advantage of the transshipment processes.

By 1870 extensive warehousing had been developed around the edges of the basin. The plan of the site that was prepared to show the proposed location of the hydraulic boat lift, indicates major warehousing facilities to the west of the basin, which extends between the river bank and the towpath of the canal. This site belonged to the Anderton Company Limited and appears to have contained two double inclines and numerous tramways. Along the northern edge of the basin there were a number of smaller warehouses with one double incline, two stages with chutes and, what appears to be, another two stages possibly incorporating cranes. Finally, to the east of the basin was more, extensive, warehousing, this time in the ownership of Traders North Staffordshire Carrying Company, presumably an offshoot of the railway company that then owned the Trent & Mersey Canal. Again, the site extended the full width of the land available between the river and the canal, and included what appear to be three separate double inclines, two of which give the impression of extending to the river's edge.

The clear impression gained of the Anderton Basin at the end of the 19th century is one of intense activity, which can best be compared to Crewe Station during the heyday of the railways or perhaps the Gravelly Hill interchange

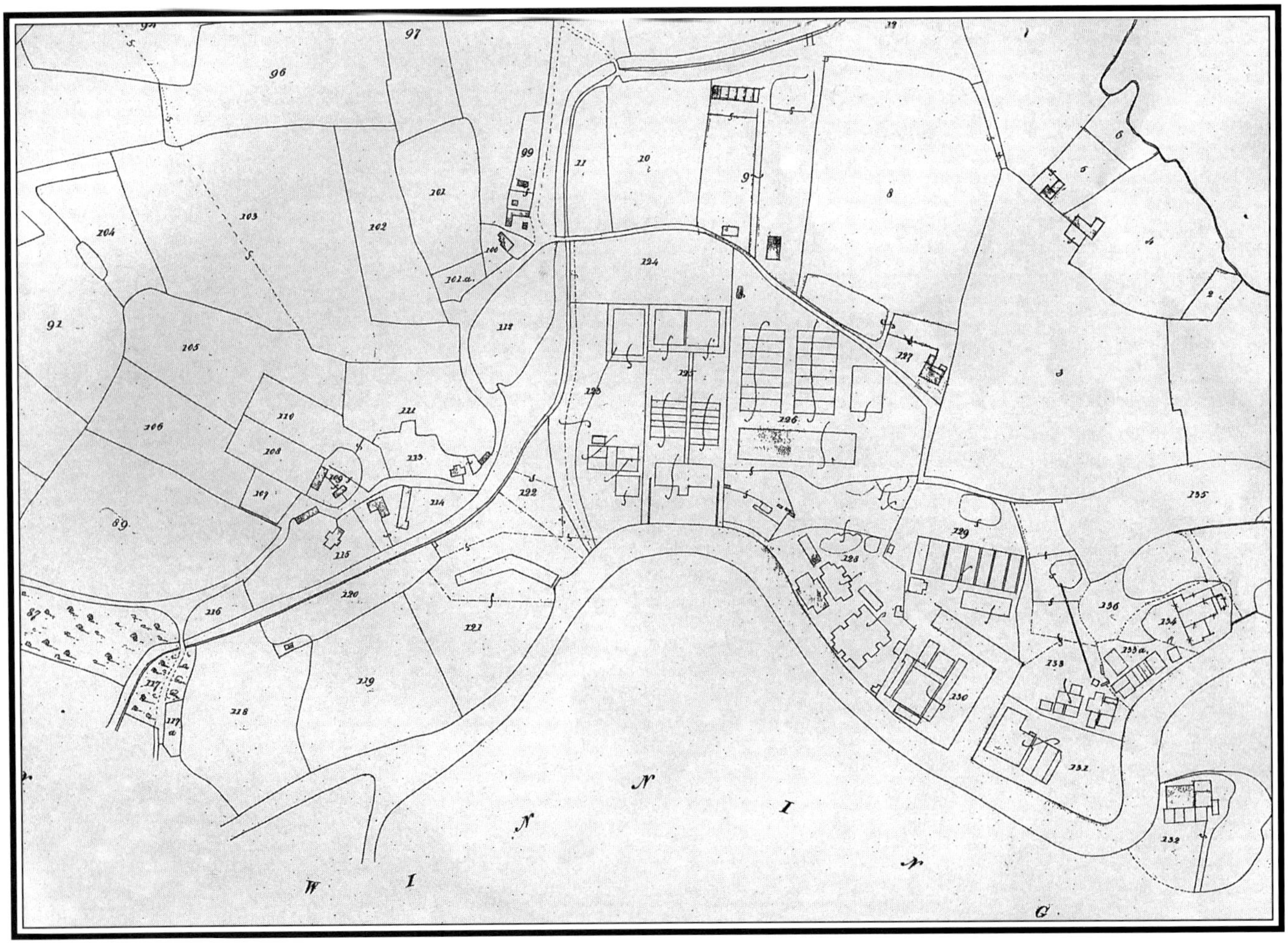

Tithe map of Anderton and Winnington, circa 1800, showing the basin and various works between the Weaver and the Trent & Mersey Canal, and rural land on the other side of the river to the south. Courtesy Cheshire Record Office, Chester

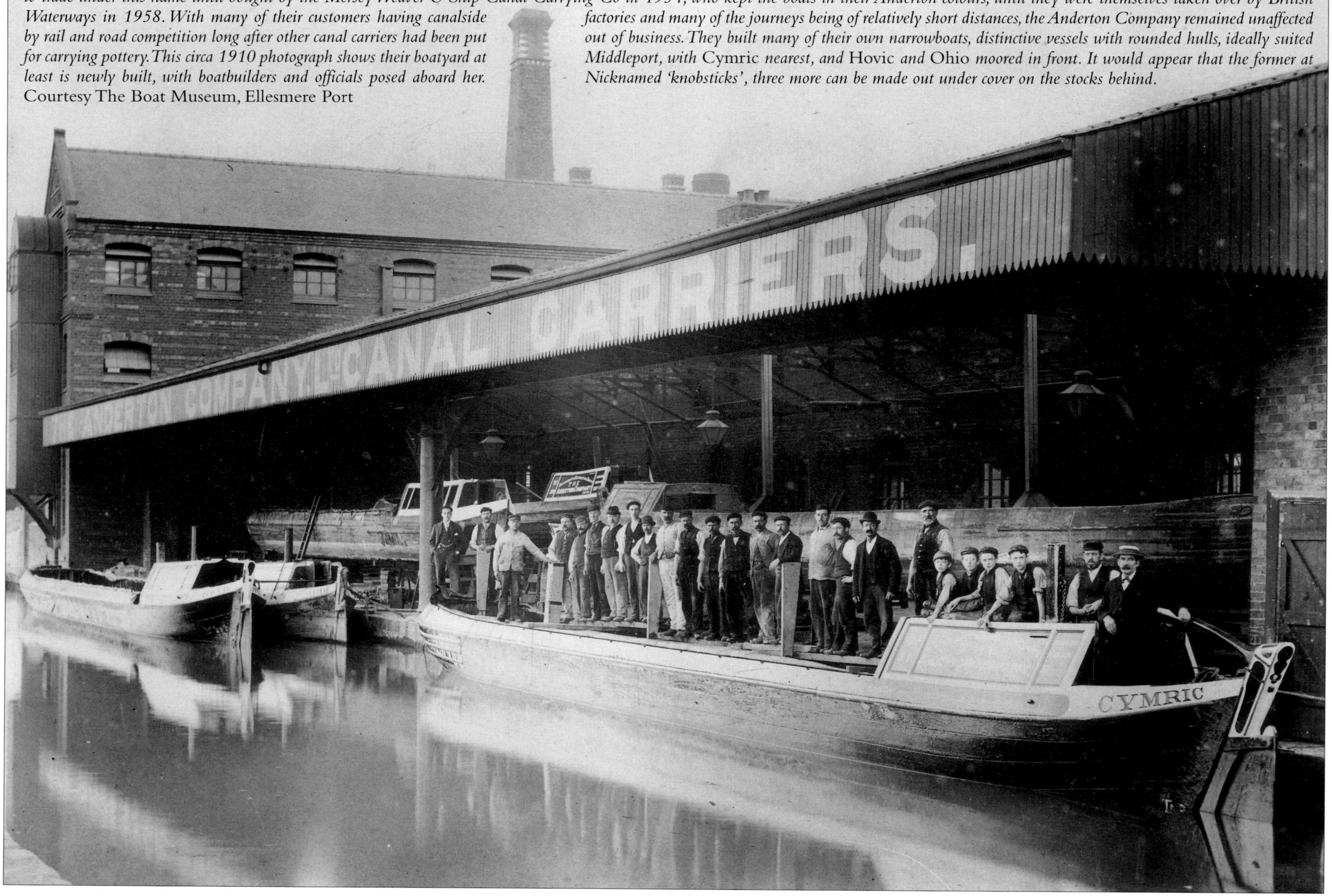

In 1836, Alexander Reid & Co changed their name to The Anderton Company, after taking over the transshipment trade at Anderton. Despite several changes of ownership, the company continued to trade under this name until bought by the Mersey Weaver & Ship Canal Carrying Co in 1954, who kept the boats in their Anderton colours, until they were themselves taken over by British Waterways in 1958. With many of their customers having canalside factories and many of the journeys being of relatively short distances, the Anderton Company remained unaffected by rail and road competition long after other canal carriers had been put out of business. They built many of their own narrowboats, distinctive vessels with rounded hulls, ideally suited for carrying pottery. This circa 1910 photograph shows their boatyard at Middleport, with Cymric *nearest, and* Hovic *and* Ohio *moored in front. It would appear that the former at least is newly built, with boatbuilders and officials posed aboard her. Nicknamed 'knobsticks', three more can be made out under cover on the stocks behind.*

Courtesy The Boat Museum, Ellesmere Port

THOMPSON

This photograph, showing salt chutes surrounding the Anderton Basin, dates from around 1900; the stairway from the upper platform to the island, removed in 1908 during the conversion works, can just be made out through the columns on the extreme right. The complex arrangement of chutes allowed several vessels to be berthed at busy times, whilst loading or waiting. Note the remains of one of the inclines still in situ running down the bank and underneath the aqueduct carrying the canal. On the left, the Weaver sailing flat Thompson *of Liverpool is loading salt; she belonged to Henry Thompson, owner of the Lion Saltworks.* Courtesy Ellesmere Port Boat Museum

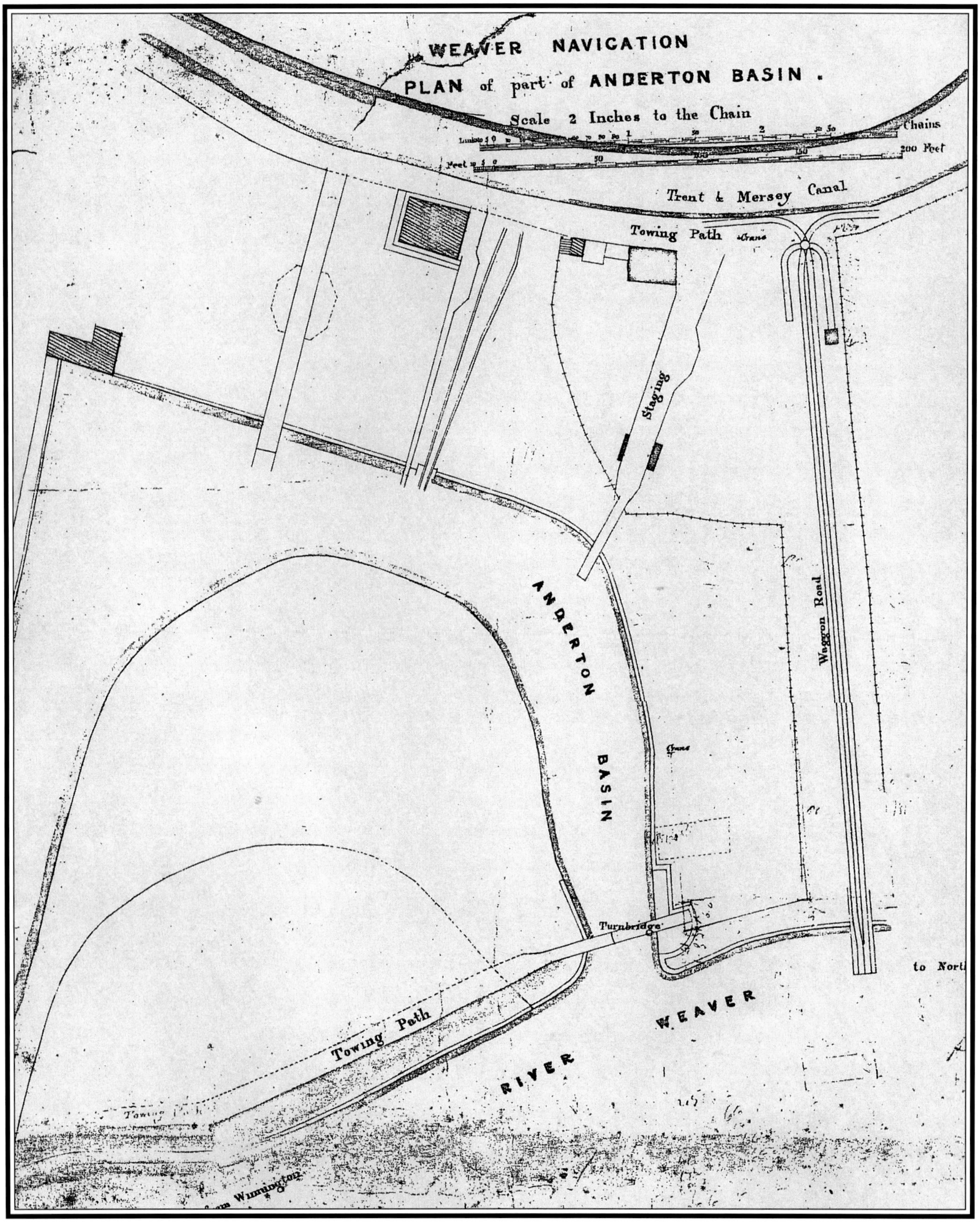

Plan of part of the Anderton Basin, undated but probably circa 1830. It may have been prepared as part of the plans for the second entrance to the basin, just shown on the left, which was constructed in 1831. At the same time a further tramroad incline was built from the canal down to where the L-shaped building is shown, at the top left-hand corner of the basin as seen here. The plan appears to show three sets of salt chutes down to the basin itself, with a self-acting tramroad incline (laden waggons going down hauling empties back up) alongside the middle chute. On the right, leading from the Trent & Mersey Canal, via a waggon turntable to a stage on the River Weaver, is another short length of tramroad. Alongside the Trent & Mersey, adjacent to the middle chute, is a stable block used for looking after the horses which brought narrowboats to this point on the canal.

Courtesy British Waterways

A busy view through a forest of ships' masts of the New Dock at Weston Point around the end of the 19th century. This became the entry point off the Mersey for smaller vessels using the Weaver and a place of transshipment for cargoes brought in on larger ships. Indeed, the Mersey flats on the far left are transferring cargo from the tramp steamer they are moored alongside, whilst behind is a barge laden with bunkering coal which the larger vessel is about to take on board. Mike Day collection

Hunt's Locks on the Weaver Navigation, about 1910. Pictured are the Salt Union packet National, *built in 1876, which has the dumb barge* Two Sisters *in tow. The latter was built in 1861 as a sailing flat. On the left is the Salt Union's inspection launch* Pacific, *whilst the Cheshire Lines Committee's viaduct carrying the Chester – Manchester railway features in the background.* Mike Day collection

A 1930s view of the steam packet Constance*, another of Henry Thompson's boats, loading salt beneath the chutes at Anderton, brought by narrowboat the short distance along the Trent & Mersey from the company's saltworks at Marston. The packet* Weaver Belle *and her consort the large dumb barge* Gowanburn*, both owned by Henry Seddon, are behind.* Courtesy The Salt Museum, Northwich

In a manner that went unchanged for decades, white salt is seen being tipped down one of the chutes at Anderton from a handcart in the 1930s. The cart would have first been filled by hand from a narrowboat on the canal. Courtesy The Salt Museum, Northwich

('Spaghetti Junction') between the M6 motorway and the Aston Expressway in Birmingham today. Is it any wonder, therefore, with this hive of activity at Anderton, that the Trustees turned their attention to the possibility of a connection between the two waterways which did not require the labour intensive, and hence, costly activity of transshipment from canal to river vessels and vice versa? It seems likely, however, that the rivalry with the Trent & Mersey Canal for a larger share of the available traffic led to the consideration of innovative solutions to the problem.

THE DEVELOPMENT OF ANDERTON BASIN 1790 - 1872

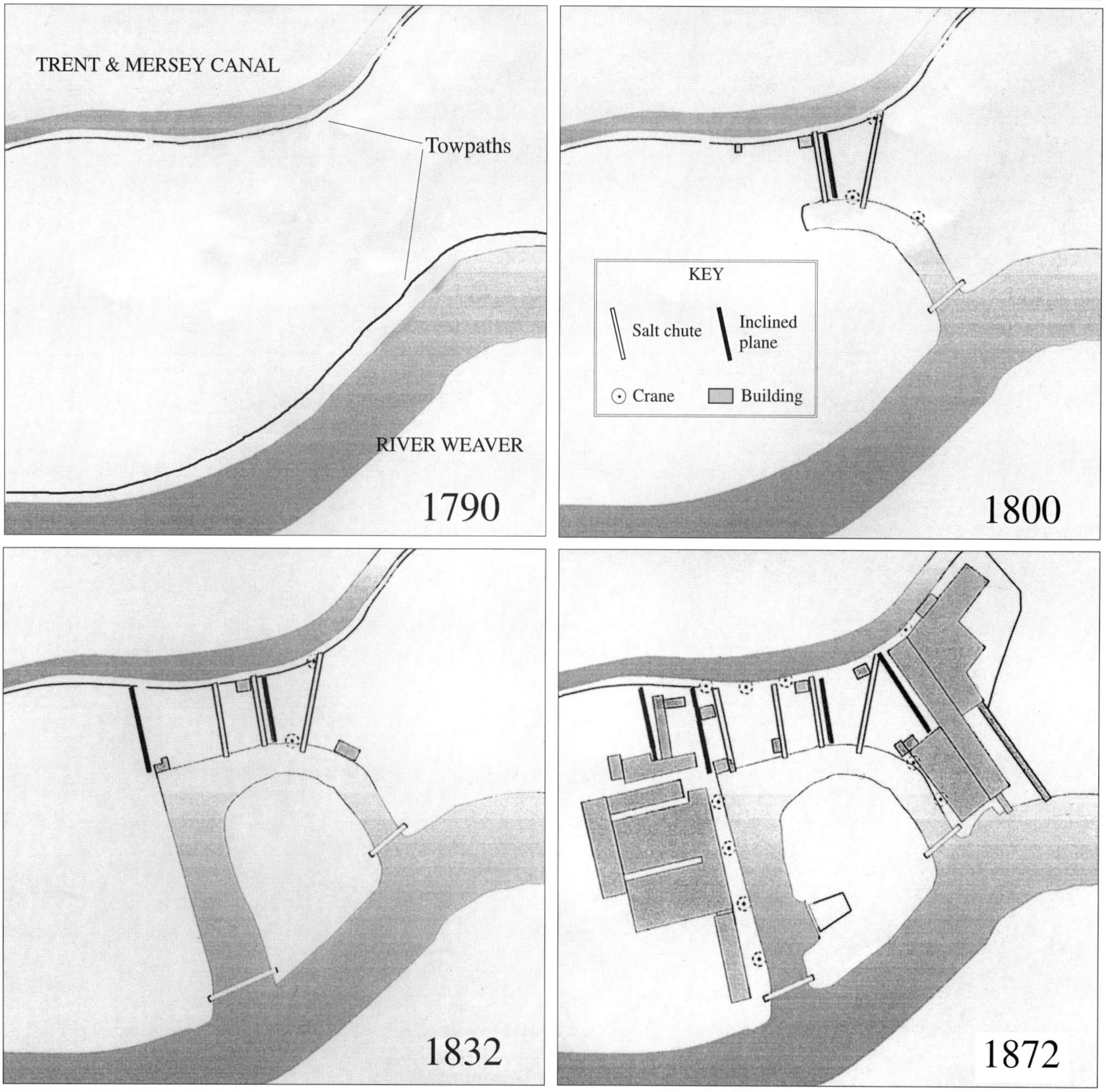

Copyright Sue Lamb/Dave Palmer 2000

INFORMATION SOURCES FOR CHAPTER 2

Notes on the River Weaver Trust, *two papers read by Thos. Ward, J.P. before the Northwich Debating Society in 1887.*
The Weaver Navigation, *by Harry Wardale.*
The Navigation of the River Weaver, *by Willan, 1951.*
Anderton Basin - An Archaeological Survey, *by Jill Collens and Stephen Penney for Cheshire County Council, October 1989.*
Knobsticks – Canal Carrying on the Northern Trent and Mersey, *by Robert J. Wilson, 1974.*
Canals Inclines and Lifts, *by David Tew. Alan Sutton Publishing, 1984.*
The Weaver Packets, *by Edward Paget-Tomlinson.* Archive *(Journal for British Industrial & Transport History)* Issue 10, June 1996.

Edwin Clark, designer of the 1875 hydraulic lift at Anderton. The painting (artist unknown) was donated to the Institution of Civil Engineers by Mess'rs Clark, Standfield & Clark. Courtesy The Institution of Civil Engineers

CHAPTER 3
DEVELOPMENT OF THE HYDRAULIC LIFT

'Our Engineer having explained the working of the proposed lift for the transhipment of traffic at Anderton and having stated that the cost of such a lift, to hold two narrow boats at once, and the basins communicating therewith would not exceed £12,000 we are of the opinion that we should lay the plans before the North Staffordshire Railway Company to ascertain whether they will concur with us in the construction of such lift and basin and we are further of the opinion that this subject should be referred to the Anderton Committee.'
Entry on 7th July 1870 in the *Weaver Minute Book 1864 - 1870*

The above statement comes from the minute book of the Weaver Trustees and could well be the first written reference to the proposal to build a boat lift at Anderton. The Engineer referred to was Edward Leader Williams, who held the post of 'Engineer to the Trust' between 1856 and 1872 (he should not be confused with Edward Leader Williams Senior, who was his father and an equally successful engineer).

The North Staffordshire Railway Company was, at that time, the owner of the Trent & Mersey Canal, having taken it over in 1845. However, unlike many similar take-overs of that time, the North Staffordshire's acquisition of the Trent & Mersey was a relatively friendly affair, with the canal being retained as a fully operational part of the company. Consequently, in view of the likely impact that a new boat lift at Anderton would have on the balance of trade between the competing waterways, it was thought prudent by the Weaver Trustees to enter into early discussions with the railway company in an attempt to minimise any objections that might be raised.

Sir Edward Leader Williams, Engineer to the Weaver Navigation Trust, 1856 - 1872.
Courtesy Manchester Ship Canal Company

The Anderton Committee were presumably a smaller grouping of the Trustees with particular responsibilities in respect to the boat lift site.

Shortly after the meeting of the Trustees in July 1870, Leader Williams met with some of the directors of the North Staffordshire Railway '*to explain the model of the lift*'. It is not clear whether this should be taken literally as a small scale physical model of it or perhaps the concept and operation of the proposed lift. Furthermore, the reaction of the directors to the presentation goes unrecorded. There is certainly no indication of the Railway Company contributing towards the cost of the construction of the lift, which presumably was one of the main purposes of the meeting. Another objective would have been to seek the support of the Company for the passage of the necessary Bill through Parliament. In this too, however, the Trustees would appear to have been unsuccessful.

* * * * * * * * * *

When built, the lift would, for the first time, enable laden boats to pass between the River Weaver and the Trent & Mersey Canal, without the need for the costly transshipment of their cargoes, principally salt and pottery goods. Hitherto, all cargoes being transferred between the two waterways had to be laboriously off-loaded from the incoming vessels and hauled overland, by one means or another, before being loaded into outgoing vessels. With the volume of traffic in existence at the time, which was primarily downhill from the canal to the river, the expenditure of time and energy associated with transshipment would have been enormous. The proposed Anderton Boat Lift was, therefore, an elegant and bold solution to the commercial problem of transshipment of goods between the two waterways.

Because of the importance given today to the Anderton Boat Lift, it is easy to fall into the way of thinking that the proposed lift was of novel design and unprecedented in engineering terms. This, however, would be a misunderstanding. There are several examples of boat lifts having been designed and constructed in Britain prior to 1870 and the hydraulic lift at Anderton had many similarities with these earlier structures. It would be wrong to assume that the one at Anderton was conceived and developed without reference to any of its less illustrious predecessors. This was confirmed six years later during a meeting at the Institution of Civil Engineers in London on 21 March 1876, which was chaired by George Robert Stephenson. Sidengham Duer, an assistant to the lift designer Edwin Clark, and Supervising Engineer of the construction works on site, presented a paper entitled *Hydraulic Canal Lift at Anderton*. In the discussions that followed the paper, Leader Williams is recorded as saying:

'There was nothing at all new in a canal lift. In the first volume of the Proceedings of the Institution, (1838), there was a paper descriptive of perpendicular lifts for passing boats on canals, as erected on the Great [sic] *Western Canal, but those only dealt with boats of 8 or 10 tons, while on the Weaver the craft were 100 tons and the height to be raised was 50 feet 4 inches A strong lift was therefore required; but it appeared worth trying, and the Weaver Trustees placed the matter in his hands.'*

It is not clear to what he was referring in the above quote, when mentioning craft of 100 tons on the Weaver; the largest boat able to be carried by the lift was a barge with a laden weight of nominally 60 tons.

Leader Williams subsequently stated that, during the initial stages of the design of the Anderton Boat Lift, he was '*fortunately induced*' to make a visit to the Victoria Docks, at Blackwall in London, to inspect a hydraulic lift graving dock in which ships were lifted clear of the water for maintenance purposes. He was obviously impressed by the ease with which the graving dock, making use of presses (hydraulic rams), could handle ships of 2,000 tons weight, and it was this that either led to, or confirmed, the idea of hydraulic operation for the boat lift at Anderton. It seems likely that Leader Williams' visit to the Victoria Docks was as a result of prompting by Edwin Clark who, during his career, had been responsible for the design of the hydraulic graving dock there.

By that time, Clark appears to have established a worldwide reputation as a designer of hydraulic machinery and was the leading partner of Mess'rs Clark, Standfield and Clark, a consultancy practice based at 6 Westminster Chambers, Victoria Street, in London. Clark was born in 1814 and lived his childhood at Great Marlow in Buckinghamshire, where his father was a manufacturer of pillow lace. After a varied upbringing in which his education was severely curtailed by his father's financial problems, Clark gained his first engineering experience in 1846, working for Robert Stephenson as designer on the Britannia Bridge. Later in his career, he held the position of Engineer to The Thames Graving Dock Company, in which role he designed, in 1857, the hydraulic graving dock at the Victoria Docks. Clark's particular specialisms lay in electrical and hydraulic engineering. It is evident that, by 1870, Edwin Clark was sufficiently respected in the engineering community to have been consulted by Leader Williams with respect to the proposal to build a hydraulic boat lift at Anderton.

What makes the Anderton Boat Lift unique is that it was the first commercially successful boat lift which, with various modifications through its life, was kept in service for over a 110 years. If ever an engineering structure could be considered to have been a success, then this was one. Whilst there have been, and still are, other successful boat lifts in the world, the Anderton Boat Lift was the first and is the only one of its kind in Britain.

★ ★ ★ ★ ★ ★ ★ ★ ★ ★

For many years prior to 1870, the volume of traffic on the two waterways and the loss of trade being caused by the delays and costs of transshipment at the Anderton Basin had been exercising the minds of the Trustees. It is almost certain that, from time to time, they commissioned their Engineers to review the available methods of transshipment. By this means the Trustees would have been kept informed of the various developments in relation to boat lifts elsewhere in the country. For example, there are records of several experimental boat lifts having been built in Britain between 1790 and 1820 but, for differing reasons, sometimes unrelated to the success or otherwise of the designs, none of these were developed beyond the trialling stages. Prior to 1800, Robert Weldon designed and tested a lift for the Shropshire Canal near Oakengates, and Edward Rowland and Exuperius Pickering carried out similar works for a lift on the Ellesmere Canal, probably at Frankton Junction. Subsequently, James Fussell designed the first lift with two counterbalanced caissons which underwent trials on the Dorset & Somerset Canal at Barrow Hill near Mells in 1800; the trials were generally successful but further development of the lift was aborted due to lack of money for the canal project as a whole. In

Right: *Plan (bottom), end (top left) and side interior (top right) elevations of one of the perpendicular lifts erected on the Grand Western Canal in Somerset, by James Green, circa 1835. Note the cradles to hold the boats were made of wood and there were vertical-lifting gates protecting against water flowing into the lift chambers at both upper and lower levels. These drawings were reproduced in an article by Green in the* Proceedings of the Institution of Civil Engineers, Vol. 2, 1838. *The arches in the masonry allowed access to the cradles for adjustments, and also for daylight to enter. Built in two stages, thirty years apart, the Grand Western Canal was beset with problems, not least caused by the seven lifts and one inclined plane on the later extension. Four of the lifts were along the principles of Dr. James Anderson, of Edinburgh, who had published a paper on the idea in 1796. The only hydraulic canal lifts operating in the country at the time, Green attempted numerous modifications in order to get them working properly. However, as financial difficulties set in and with through navigation still not possible, he was dismissed by the canal company in 1836 before he'd had chance to complete his work.* Courtesy Ellesmere Port Boat Museum

PLANS OF BOAT LIFT BY JAMES GREEN FOR THE GRAND WESTERN CANAL 1835

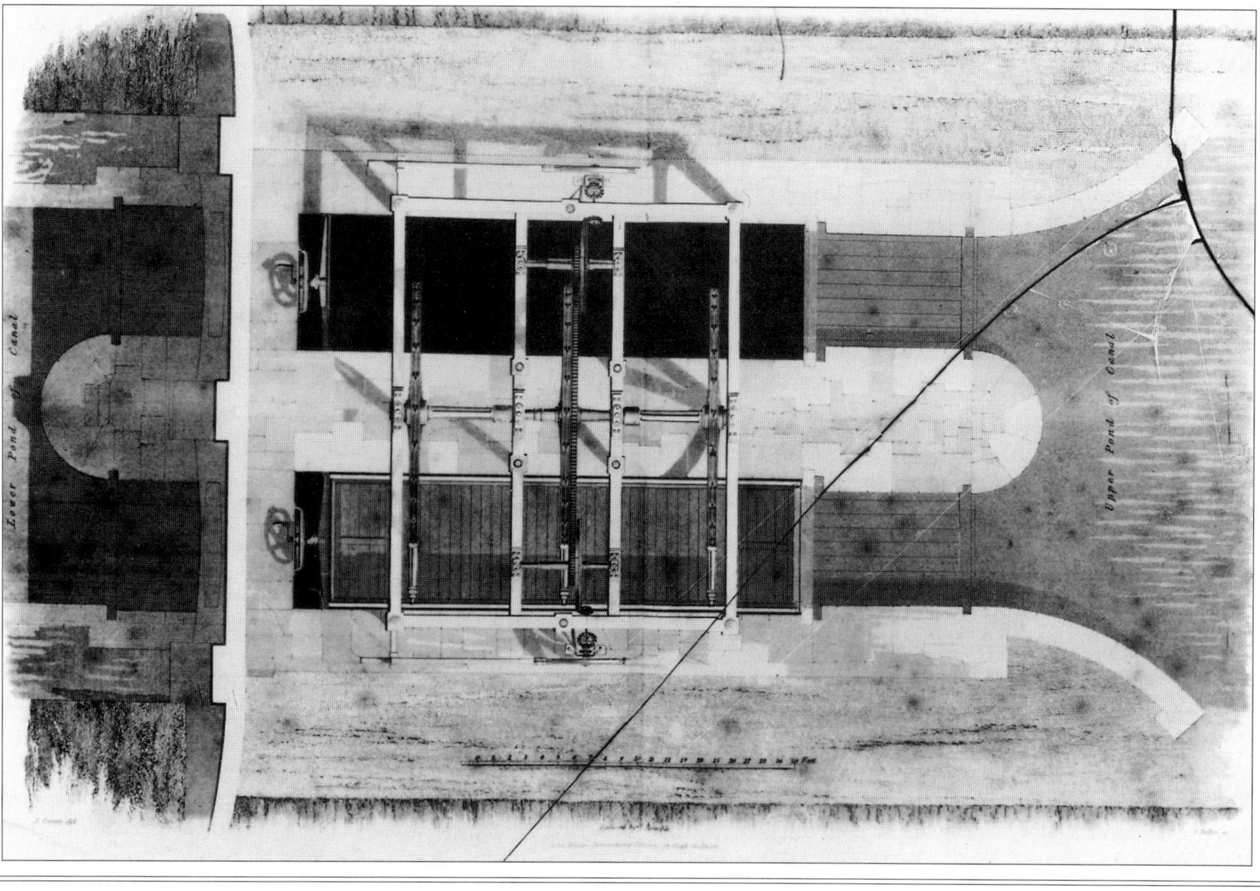

1808, a boat lift designed by Robert Woodhouse was constructed on the site of the top lock at Tardebigge on the Worcester & Birmingham Canal but again, for various reasons, the lift did not get beyond the trialling stages, eventually being replaced by a deep lock in 1815. In 1813, Colonel Sir William Congreve patented a design of a boat lift for the Regent's Canal, at Hampstead Road in London, but again there is no record of this having been built.

In the 1830s, James Green was the architect of a series of seven, relatively successful, boat lifts on the eastern section of the Grand Western Canal, between Taunton and Lowdwells in Somerset (presumably the boat lifts to which Leader Williams referred during the meeting at the Institution of Civil Engineers on 21 March 1876). These lifts, which owed much to the earlier experimental lift by James Fussell, consisted of two counterbalanced cradles within which boats could float; the cradles being in adjacent masonry chambers and interlinked by chains. The lift was set in motion by increasing the depth of water in the upper cradle by 2 inches or so. The seven lifts were built over a period of about five years and were used to lift the Grand Western tub-boats, vessels 26 feet long by 6 feet 6 inches wide, with a draught of 2 feet 3 inches when loaded. The lifts were in operation for about thirty years, until the demise of the canal in 1867 due to competition from the Bristol & Exeter Railway. They were eventually dismantled.

* * * * * * * * * *

Another aspect to the background, and eventual development of the Anderton Boat Lift which should not be overlooked, is that most of the early attempts at boat lifts described above took place against the, almost forgotten, backdrop of the rapidly expanding use of hydraulic machinery in Britain. Under the influence of pioneers such as Joseph Bramah (1748 - 1814) and William Armstrong (1810 - 1900), it was hydraulic machinery which was the driving force behind the industrial revolution. Hydraulic energy was of similar importance in the 19th century as electricity, gas or oil is in today's world, and it was only in the early years of this century that hydraulic power began to be eclipsed by these other forms of energy.

The potential of using water as a means of transmitting energy was first realised in 1795, when a patent to that effect was granted to Joseph Bramah. He subsequently invented the hydraulic press, a relatively simple machine that was, in various forms, to revolutionise Britain's industry. The hydraulic press made use of the paradox of 'the equilibrium of liquids', as originally propounded in 1647 by the French physicist Blaise Pascal, to develop very high forces by what appeared to be the input of relatively low energy; hence the paradox (the simplest modern day example of the hydraulic press is the bottle jack, which mechanics still use to raise vehicles many times their weight by means of a simple hand-lever). The hydraulic press was initially used to pack light, bulky goods, such as cotton, into dense bales, thereby enabling significant reductions in transportation costs. Prior to the invention of the hydraulic press, packing of materials was carried out by means of screw presses, which had an effective maximum capacity of nominally 25 tons. Bramah's hydraulic press, however, could in theory be built to any capacity and as early as 1799 a 400 ton press had been manufactured. During subsequent years the use of hydraulic power became a major factor in the industrialisation of Britain.

William Armstrong was one of the first to realise the full potential of hydraulic power, in the form of hydraulic rams, and one of his many achievements was to design and build hydraulically operated cranes, which were to have far reaching effects in Britain's docks and railway yards during the second half of the 19th century. He also recognised that hydraulic power could be stored by means of water towers and could be transmitted from one place to another by means of pipes. Armstrong subsequently invented the accumulator, a relatively simple – and hence cheap – device for storing hydraulic energy, which obviated the need for water towers.

In subsequent years, hydraulically driven machines were developed for many diverse uses. For example, hydraulic power was exploited by prominent civil engineers, such as Robert Stephenson who, in 1851, used rams, with a total capacity of 2,620 tons, to raise the spans of the Britannia Bridge into their final position across the Menai Straits in North Wales (it is of interest to note that Edwin Clark was Resident Engineer on site for these bridge works). Prior to the invention of the hydraulic press by Bramah, this bridge, and many other structures that benefited from the use of hydraulic machinery, could not have been built.

One other development in the use of hydraulic power of relevance to the Anderton Boat Lift, was the invention in 1866, by Ralph Tweddell, of the hydraulic riveter. This greatly simplified, and hence reduced the cost of, the erection of iron structures.

* * * * * * * * * *.

The idea of a vertical boat lift at Anderton was selected only after due consideration of various alternatives. A paper written by John Saner, the Engineer to the Trust in 1908, which was published in the 14 July 1908 edition of *The Engineer*, refers back to those early days in the conception of the lift as follows:

'The first means which suggested themselves for this purpose were the lowering and raising of vessels by a series of locks; but owing to the difference in levels of the canal and the river at this point, the great space required for such a scheme, coupled with the delay which would be involved, and the loss of water from the canal, the scheme was found impracticable. Mr. (now Sir) Edward Leader Williams, who was at that time engineer to the Weaver Trust, conceived the idea of lifting and lowering vessels by hydraulic presses, without taking any water from the canal. It should be pointed out that the canal, which for some miles runs parallel to the river, is on top of a bank, while the river is below, the difference between the levels of the water being 50ft 4in. After consultation with Mr. Edwin Clark, M. Inst. C.E., it was decided that the vessels should be lifted while floating in a trough of water, to obviate all danger of straining the hulls in the event of the cargoes

shifting. To do this involved lifting the weight of the water in the trough as well as the barges, and it was decided to have two equal troughs, so as to balance each other, and so reduce the amount of power required.'

A paper written in 1878 by Sidengham Duer, sheds more light on the reasons for the selection of a vertical boat lift as opposed to a flight of locks:

'It was constructed instead of a chain of locks as being very much cheaper both in first cost and working expenses; besides which it takes from the canal only about one per cent of the water that would be required by barges passing through a chain of locks of the same total height, and by this means fully ten times the number of barges can pass from the canal to the river or vice versa in a given time.'

One of the main problems which canal engineers faced, was that of establishing a reliable and plentiful source of supply to replenish the water in the canal lost due to evaporation, leakage and lockaging; the latter being the inevitable consequence of the use of locks to overcome changes in ground levels. Each time that a vessel passes through a lock, a significant volume of water is transferred from the upper section of the canal to the lower. A locking operation could result in the loss of as much as 50,000 gallons of water from the upper level, although there have been several ingenious methods of achieving reductions in this amount.

Bingley Five Rise Locks – also one of the 'Seven Wonders of the Waterways' – on the Leeds & Liverpool Canal are still in regular use by pleasure boaters today. Author

The total water lost from the upper section of a canal, which is not just related to the size of the lock but is also a function of the volume of traffic, needs to be replaced if adequate draught is to be maintained. A flight of locks at Anderton, an interconnection of two waterways which were heavily trafficked in the late 1800s and early 1900s, would, almost certainly, have had severe effects on water depths in the Trent & Mersey Canal, particularly during the summer months.

Furthermore, the difference in levels between the two waterways at Anderton would have required a flight of four or five locks with an overall length of several hundred feet. It was concluded, presumably by Leader Williams, that a flight of locks could not have been easily accommodated at Anderton; hence the attraction of a boat lift. It is interesting to note that the Bingley Five Rise Locks on the Leeds & Liverpool Canal, which as staircase locks make the most efficient use of land, achieve a total lift of 60 feet over a total length of some 350 feet.

★ ★ ★ ★ ★ ★ ★ ★ ★ ★

It was against this background that on 12 October 1871, a Special General Meeting of the Trustees took place at Northwich:

'to consider the desirability of constructing a lift with basins and all other requisite works for the interchange of traffic between the River Weaver and the North Staffordshire Canal at Anderton and of applying to Parliament for an Act to authorise the construction of such works or of making such other order as may be deemed to be expedient for that purpose.'

The Trustees agreed to apply to Westminster accordingly and soon after, on 21 November 1871, an article in the *London Gazette* announced the Trust's intention of constructing a boat lift at Anderton as follows:

'The Trustees of the River Weaver Navigation intend to apply to Parliament in the next session thereof, for leave to introduce a Bill for the following, or some of the following, among other powers and purposes:

(1) The facilitating of communication between the west side of the Anderton-basin of the Trustees and the Trent and Mersey Canal by means (among other things) of a cut and aqueduct and basin for the purpose of interchanging cargoes in boats and other vessels, in bulk, between the Weaver Navigation and the said canal.'

The article went on to describe the elements of the proposed lift in the following terms:

'(a) A cut commencing on land of the Trustees on the east bank of the west side of the Anderton-basin at a point about 55 yards measured northward of the River Weaver along the centre of the western arm of the said basin.

(b) An aqueduct carried across the north-eastern part of the

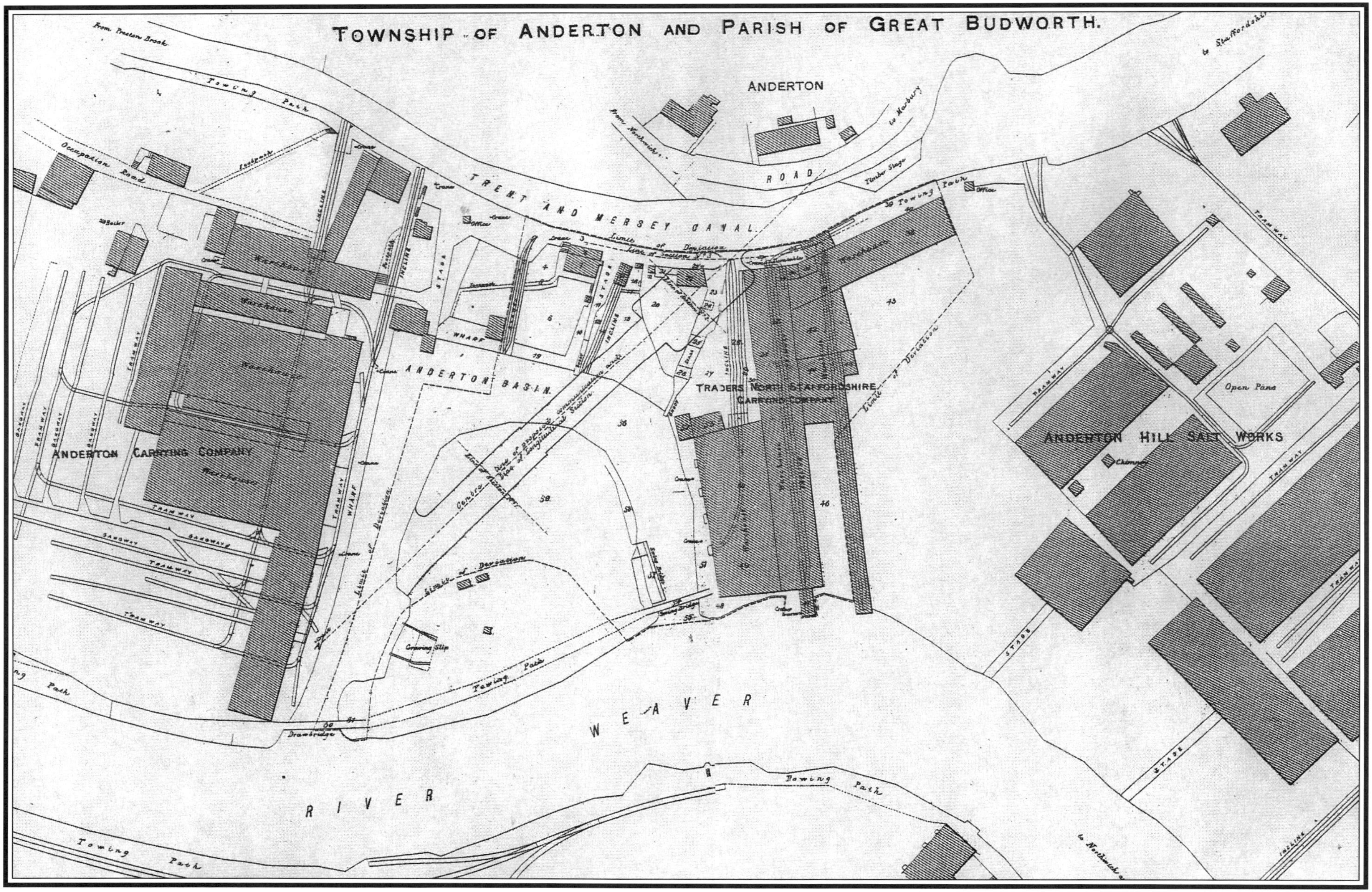

Plan of the Anderton Basin, with the outline of the proposed boat lift, as published in the London Gazette *in November 1871 to coincide with the introduction of the Bill in the Houses of Parliament. Notice the extensive warehousing, chutes, inclines and tramroads.*
Courtesy Cheshire Record Office, Chester

said basin.

(c) A side-dock or enlarged cut on lands belonging to the Trustees between the north-eastern side of the said basin and the Trent and Mersey canal.

(d) A communication or cut from the said side-dock into the Trent and Mersey Canal and through the towing path thereof on the south side of the same canal at or near a spot 40 yards north-eastward from the house belonging to the River Weaver Trustees, and occupied by Thomas Griffiths, and forming part of the premises known as the Anderton Inclines and Salt-shoots.

(e) All necessary basins, stop-gates, lifts, inclines, sluices, bridges, aqueducts and culverts for the purposes of the said communication.'

Intriguingly, this extract from the *London Gazette* makes no direct reference to a vertical hydraulic boat lift. Indeed, it almost fails to make any reference to the lift at all. One is led to speculate whether, at the time of going to press, the Trustees had decided what form the connection between the canal and river should actually take.

The years that followed the Trustees' decision to proceed would have seen frenetic activity on the detailed design of the boat lift. As already stated, and contrary to many other publications which attribute the design to Leader Williams, the principal designer of the boat lift was Edwin Clark, although he would presumably have been assisted by a team of engineers, which could well have included Sidengham Duer.

Leader Williams can, more correctly in the opinion of the present author, be attributed with conceiving the idea of a boat lift at Anderton and perhaps for the preliminary design work. Presumably, he fulfilled a supervisory role during the main part of the detailed design process and would have kept the Trustees properly informed as work progressed. It is almost certain that, as with nearly all engineering projects, the detailed design process would have resulted in a number of changes to Leader Williams' original concept, with consequentially higher costs, but more of this later.

At the previously mentioned meeting at the Institution of Civil Engineers in London on 21 March 1876, Duer provided some detail on the early design process:

'It was thought advisable that the barges should be lifted while floating in a trough of water, to obviate all danger of straining, in the event of their cargoes shifting. This necessitated lifting the weight of the water in the trough as well as the barges, and was found to require a large steam-engine. It was then proposed to make the lift double, or to have two equal troughs which, being by some means connected, should counterbalance each other, like a pair of scales, when only one-half of the engine-power would be required. It was further improved by the idea of taking a little water into the upper trough from the upper level, so as to make it heavier than the lower one, and that in its descent it should do nearly all the work of lifting the other trough.'

During the discussion time at the end of the meeting, Leader Williams explained that, prior to involving Clark and Duer on the project, he had designed a lift with a system of counterbalance weights with water, '*which undoubtedly was cumbersome*'. Afterwards, however, the option of two presses per caisson, one at each end, was considered. This option though, posed problems with respect to synchronisation of the operation of the presses, with the risk of a caisson moving out of horizontal alignment and the subsequent movement of water to one end throwing more load onto the shorter ram. As a consequence of this, the decision was taken to adopt a single, 3 foot diameter, ram, centrally located under each caisson. At the meeting, Duer subsequently added that another early proposal was for four presses per caisson; that is to say one in each corner. Again, however, this option was rejected due to the difficulty in keeping the caissons horizontal.

At the meeting Clark added that, with respect to the single ram option:

'The lifting of so large a pontoon with a single press was a problem that had involved a good deal of thought. It was a large press, with a long stroke, and it was a question how far, with so heavy an overhanging weight, reliance could with safety be placed on a bracket of that kind. But inasmuch as the load was immovable, the water being quiescent, it appeared to be a case in which there was no danger in putting on top of the single press a weight which, with any varying load, would have involved danger. He was convinced from the experience on the Weaver, that the limit to which a single press might be applied had not been attained.'

★ ★ ★ ★ ★ ★ ★ ★ ★ ★

On 18 July 1872, Royal Assent was granted for the Weaver Navigation 1872 Act (the eighth Weaver Act), which provided the legal mechanism for the construction of the boat lift. Soon after, on 8 August 1872, the minutes of the Trust record that:

'Mess'rs Emmerson and Co. delivered in their revised estimate of the proposed lift at Anderton which they had prepared by the desire of our Engineer, amounting to £23,420. We are of the opinion that a contract should be entered into with Mess'rs Emmerson and Co. upon the terms of this letter.'

At the same time, Leader Williams advised the Trustees of his imminent resignation to take up the post of Engineer to the nearby Bridgewater Canal Company (later still, he became Engineer to the Manchester Ship Canal). Two months after Leader Williams' resignation, John Watt Sandeman was appointed as the new Engineer, with Leader Williams being retained by the Trustees as Consulting Engineer until the completion of the boat lift works. Sandeman's starting salary was £800 per annum, with a house.

A question which arises from the above extract is how and why Emmerson & Co. were selected. One possibility is that Leader Williams was already familiar with the company, perhaps because it had carried out work previously for the Trust on the Anderton Basin or elsewhere, and thus he was satisfied that the company had the ability and resources to fabricate and erect the boat lift. Whatever, there is no suggestion in the records that any competitive tendering for the works took place.

THE CANAL DU CENTRE AT LA LOUVIERE IN BELGIUM

Top: *Lift No.1.* ***Bottom:*** *Lift No.3. Two of the four boat lifts still in use on this canal in 1999.* All photos author

A circa 1900 postcard showing the hydraulic boat lift on the Neufosse Canal at Les Fontinettes in Northern France. Opened in 1888, it was built to accommodate the wide barges prevalent on European waterways. Author's collection

The minutes of the Trustees record on 2 September 1872 that:

'Our Engineer reported that he had been satisfied by Mess'rs Emmerson as to the correctness of the revised estimate for the lift and he produced the Designs and Specifications and our Clerk having laid before us the draft of the proposed Contract we directed him to proceed with such modifications as may be necessary.'

Within a few days the Contract for the fabrication and erection of the 'Patent Hydraulic Lift at Anderton' was awarded to Emmerson Murgatroyd & Co. Ltd, of Stockport and Liverpool, at the cost of £28,420, a sum significantly higher than originally envisaged.

There would appear to have been several reasons for the twofold increase in cost from Leader Williams' original estimate of £12,000 in July 1870, to the eventual Contract Sum, just over two years later, of £28,420. The most significant of these appears to have been the higher than expected cost of the ironwork. Again at the meeting at the Institution of Civil Engineers in March 1876, Sandeman advised that:

'about September 1872, when the contract for the ironwork, &c., was let, the price of iron had reached the highest point ever known in this country.'

No doubt, however, as with all construction projects, the cost of the works at the beginning was probably seriously under-estimated; an error, caused by natural optimism, that practising engineers tend to make even today.

★ ★ ★ ★ ★ ★ ★ ★ ★ ★

Finally, it is worth recording that, on the completion of the Anderton Boat Lift, Edwin Clark became involved in the design of a number of boat lifts on the Continent. These made use of the same hydraulic operating principles employed at Anderton but contained significant improvements based on the practical experience he had gained there.

On the Neufosse Canal at Les Fontinettes in northern France, a boat lift was built to parallel an existing flight of five locks which were a bottleneck to canal traffic. This lift was larger than that at Anderton and was capable of handling 300 ton barges. Unlike Anderton, the lift contained a dry well and consequently did not require an accumulator to overcome buoyancy effects. In addition, the rams were 2 metres in diameter and the presses were fabricated from steel with copper linings. The structure lifted boats vertically over a height of nominally 13 metres. The lift at Les Fontinettes was officially opened on 8 July 1888 and remained in service into the 1950s, when it was replaced by a single deep lock needed to accommodate larger boats. The lift has been retained by the French authorities as a static monument.

Four similar boat lifts were constructed on the Canal du Centre at La Louviere in Belgium, nominally 50 miles south of Brussels. These lifts were designed to carry boats of between 300 and 400 tons and also made use of dry wells and steel presses. The first of the Belgian lifts was designed by M. P. Nolet, an engineer employed by Clark, Standfield & Clark, and was opened on 4 June 1888 by King Leopold II. For various reasons, the other lifts were only completed in 1917, after Clark's death in 1894. The total lift of the four structures is nominally 68 metres. All four are still in operational use today, although a single, replacement lift is currently under construction.

INFORMATION SOURCES FOR CHAPTER 3

The Weaver Minute Book - 1864 to 1870. *Held at the Cheshire Records Office.*
Selected Weaver Trustee Minutes - 1865 to 1947. *Jack Horton, 1986.*
The Hydraulic Canal Lift at Anderton on the River Weaver. *Paper by Sidengham Duer, ICE Proceedings Vol. 45, pp 107-129*
Obituary to Edwin Clark. *Proceedings of the Institution of Civil Engineers.*
Canals Inclines and Lifts, *by David Tew. Alan Sutton Publishing, 1984.*
Hydraulic Machines *(Shire Album No. 144) by Adrian Jarvis. Shire Publications, 1985.*
The Grand Western Canal, *by Helen Harris. David & Charles, 1973.*
The Illustrated History of Canal and River Navigations, *by Edward Paget-Tomlinson. Sheffield Academic Press, 1993.*
The Anderton Boat Lift. *Paper by J. A. Saner on behalf of the Weaver Trustees, The Engineer 14 July 1908.*
Hydraulic Lift Lock at Anderton on the River Weaver in Cheshire. *Paper by Sidengham Duer, December 1878.*
Article in the London Gazette, *21 November 1871*

Aerial view of the River Weaver as it passes through the centre of Northwich, taken from the top of the tall gas holder seen in the background of the top picture, page 147. The view is undated but is probably in the 1930s; Hayhurst Yard is in the foreground (it was later extended and a larger repair shed built under British Waterways) and the Weaver Navigation Yard in the background, behind Hayhurst swingbridge. Courtesy The Boat Museum, Ellesmere Port

CHAPTER 4
CONSTRUCTION OF THE HYDRAULIC LIFT

'The work was executed in the most satisfactory manner by Mess'rs. Emmerson & Murgatroyd, of Stockport, from the designs and under the superintendence of Mr. E. Leader Williams, then Engineer to the River Weaver Navigation.'

Extract from paper by Lionel B. Wells to the Manchester Association of Employers, Foremen, and Draughtsmen, entitled *Visit to Northwich, the Hydraulic River and Canal Boat Lift at Anderton and the New Flood Sluices and Locks at Dutton*, dated 17 May 1884

The above extract from a paper given by Lionel B. Wells, who succeeded Sandeman as Engineer to the Trust on 5 March 1877, is not entirely accurate. Tactfully perhaps, it credits Edward Leader Williams with the '*designs*' and '*superintendence*'; however, whilst as stated in the previous chapter, Leader Williams may have had the original idea of a hydraulic boat lift at Anderton and subsequently undertook preliminary design work, it was in fact Edwin Clark who carried out the detailed design of the superstructure, with assistance, so it would seem, from Sandeman on the foundations. Furthermore it is Sandeman, and Edwin Clark's Resident Engineer, Sidengham Duer, who should be credited with the supervision of the works on site. At that time, Leader Williams was acting as Consulting Engineer to the Trust. This situation is confirmed by an extract from a hand-written report by Clark on 18 April 1882 which states that the Weaver Trustees had:

'at the suggestion of their then Engineer Mr. E. L. Williams, decided on forming a means of communication between the river Weaver and the Trent and Mersey Canal at Anderton, and as I was then engaged on my system of Hydraulic Docks as employed at the Victoria Graving Dock and at Bombay and Malta, the matter was placed in my hands.

Having prepared the designs, which I subsequently patented, I appointed one of my assistants, Mr. Sidengham Duer, my resident Engineer for this work, and for the inspection of the materials.'

Clark's role as main designer appears to be corroborated by an entry in the minutes of the Weaver Trust on 1 March 1875, which states:

'Mr. Blake having read the correspondence with reference to the claim of Mr. Edwin Clark for a certain percentage on the cost of the Anderton Lift, we are of the opinion that we should pay Mr. Clark 3 per cent of the total amount of the Contract for the works being one half of the percentage claimed, the contractor paying the other half.'

On this basis, Clark was rewarded handsomely for his efforts. Six per cent of the final cost of the ironwork superstructure is approximately £1,770, which, at the time, was a not insignificant sum.

* * * * * * * * * *

As for the new Engineer to the Trust, Sandeman, it would appear that he was not slow in getting involved with the design process for the new lift. Seemingly, soon after the award of the Contract to Emmerson Murgatroyd & Co., he saw fit to amend the plans in order to strengthen the foundations of the structure, presumably to take account of the high risk of ground subsidence in the area. Much of the Cheshire Plain was, and to a lesser extent still is, prone to ground subsidence due to the collapse of underground caverns and tunnels created by many years of brine pumping. The town of Northwich and its surrounding villages were particularly vulnerable to this problem. Evidently Sandeman considered that the original design for the foundations of the proposed boat lift had not taken sufficient cognizance of this.

The minutes of the Trustees on 6 January 1873 record another change to the original design brought about by their new Engineer:

'Our Engineer having explained the alteration which he proposed to make in the direction of the lift at Anderton, we approve of the place as amended.'

It is not particularly clear from the records what the precise nature of the change in '*direction of the lift*' amounted to and why it was necessary. One can only conclude that Sandeman saw some benefit in a realignment of the structure from that originally proposed by Leader Williams or Clark.

Apart from the above snippets, there are few references in the available records to shed further light on the design process.

* * * * * * * * * *

The same can also be said with respect to the construction

Photographs of the construction of the hydraulic boat lift at Anderton are rare. This view, taken probably in 1874, shows work under the auspices of the contractors Emmerson, Murgatroyd & Co nearing completion. The cylindrical accumulator is clearly visible to the left of the caissons, which are themselves seen in clear detail, as are the rams which lifted them up and down. Courtesy Merseyside Maritime Museum, Liverpool

process. By all accounts it would appear that the fabrication and construction works were carried out over a period of some 30 months between the end of 1872 and July 1875. The construction would have required extensive temporary works in the form of timber scaffolding, with numerous hoists and fixed cranes. Even with the plant and equipment available today, the erection of the boat lift would be a relatively major undertaking.

It is evident that Sandeman was very closely involved in the construction of the structure's foundations on site, presumably alongside Duer. Some interesting aspects of their construction come out of the proceedings of the meeting at the Institution of Civil Engineers on 21 March 1876, in which a paper was read by Duer on the boat lift. The minutes of the proceeding record Sandeman stated:

'In carrying out the foundations of the canal lift, and the masonry of the basins and approaches, the design and execution of which devolved upon him, the following features might be interesting. The foundations for the press cylinders, which contained the lifting rams, were provided by sinking cast-iron cylinders of 5 feet 6 inches in diameter to a depth of 70 feet. The ground in the interior was excavated by hand, and the water excluded by compressed air, which at one time reached 26 lbs. per square inch above atmospheric pressure. The pneumatic process had to be resorted to, in consequence of the great infiltration of water from the river, through the strata of sand and gravel, which extended to a depth of upwards of 50 feet; but below this, the hard stratified marl, which overlaid all the rock salt in this district, was reached, and this formed an excellent foundation for the press cylinders.'

He goes on to talk about the foundations to the canal basin as follows:

'The piling for the support of the retaining wall for the upper basin had to be of a substantial character, in consequence of a portion of the superstructure having to be built upon

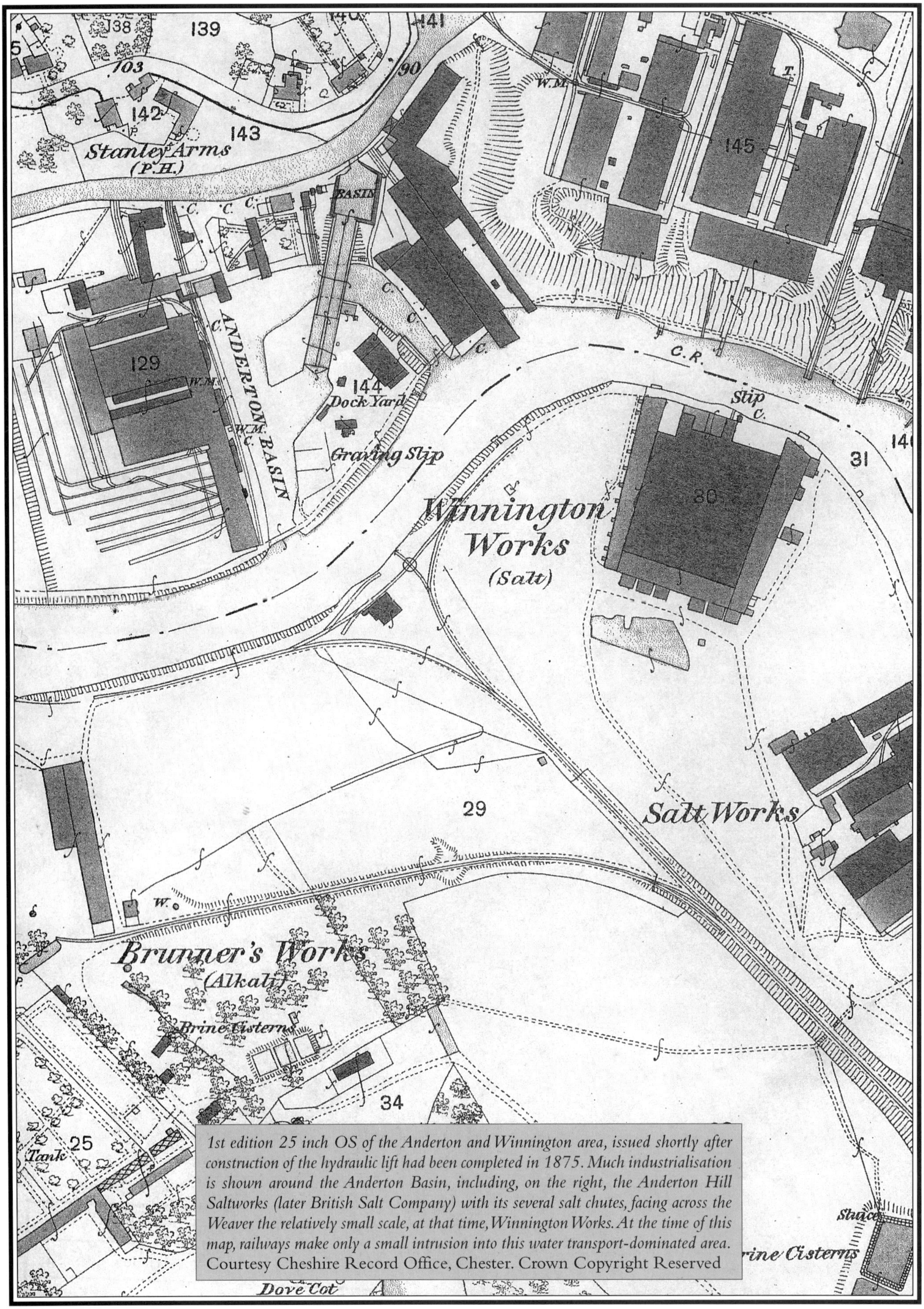

1st edition 25 inch OS of the Anderton and Winnington area, issued shortly after construction of the hydraulic lift had been completed in 1875. Much industrialisation is shown around the Anderton Basin, including, on the right, the Anderton Hill Saltworks (later British Salt Company) with its several salt chutes, facing across the Weaver the relatively small scale, at that time, Winnington Works. At the time of this map, railways make only a small intrusion into this water transport-dominated area.
Courtesy Cheshire Record Office, Chester. Crown Copyright Reserved

the face of a bank composed of sand and clay. The bottom of the upper basin was puddled clay, about 1 yard in thickness, and below the level of this clay apertures were left through the masonry of the retaining walls, so as to prevent any accumulation of water behind by infiltration from above; and although slight leakages showed themselves at first, through these apertures, they soon ceased, and therefore proved that there was no further risk of water permeating through the clay bottom of the basin.'

The cast and wrought iron elements of the boat lift and aqueduct superstructure were fabricated by Emmerson Murgatroyd & Co, most probably at their yard in Stockport. Smaller components from Stockport could either have been brought overland to Anderton by horse and cart or by barge via the Ashton, Bridgewater and Trent & Mersey Canals. However, the transportation of the larger components, such as the cast iron columns for the lift, would have posed major difficulties, especially from Stockport. It is therefore possible that the larger elements were fabricated at the company's Liverpool premises, in order to take advantage of the better transport route to Anderton via the River Mersey and the Weaver Navigation.

Work on the structure must have started some time between September 1872 and March 1873, because the minutes of the meeting of the Weaver Trustees on 3 March 1873 record that:

'Our Engineer delivered in his Certificate for the payment of £4,400, the first instalment for the Anderton Lift.'

Evidently, by reference to the minutes of the meeting of the Weaver Trustees on 7 December 1874, the fabrication stage of the works did not run totally smoothly with numerous design changes having to be made, presumably under the direction of Duer, the Supervising Engineer for the Works. The minutes record that:

'Mess'rs Emmerson Murgatroyd and Co. have applied to be paid extras on the Contract for the construction of this lift on the grounds that the weight of iron which has had to be used has much exceeded the quantity upon which the Contract was based; and it appearing from the report of our Engineer that numerous alterations have been made with his approval during the progress of the work as causing heavy additional expenditure and that such alterations and the extras, consequent upon the additional quantity of materials now amount to the sum of £2,882. We are of the opinion that such sum should be paid.'

In responding to a question at the end of the meeting at the Institution of Civil Engineers on 21 March 1876, Sandeman advised that:

'the prime cost of the ironwork and machinery, including the royalty, of the hydraulic canal lift at Anderton was £29,463, while that of the foundations of the lift, masonry of basins, approaches, &c., was £18,965, making a total of £48,428.'

The final cost of the whole of the works is confirmed in the Trust's *Summary of Year's Business* in August 1882. One interpretation of the records quoted above is that the foundations of the lift were constructed by the workforce of the Weaver Navigation Trust, under the direct control of Sandeman, whilst the above ground works were carried out by Emmerson Murgatroyd under the supervision of Duer. If this was so, then the final cost of the ironwork of £29,463 should properly be compared to Emmerson Murgatroyd's original contract sum of £28,420, which does not seem to be an excessive overspend. However, this does not seem to be supported by the decision of the Trust in December 1874 to make extra payment to the contractor of £2,882. A possible explanation is that the contract for the whole of the works was awarded to Emmerson Murgatroyd & Co. and that they subsequently sub-let the foundation works to the workforce of the Weaver Navigation. If the latter supposition is correct, then the original contract sum of £28,420 should be compared to the final cost of £48,428, an over-run of almost £20,000. This would then explain why so much time was given over to discussing the costs of the project at the meeting at the Institution of Civil Engineers on 21 March 1876.

Whatever explanation is correct, the bulk of the increased cost of the works during the construction stage was undoubtedly due both to the need to strengthen the foundations and the increased weight of the ironwork to the superstructure.

Notwithstanding the significant overspend, Leader Williams was of the opinion that the overall cost of the boat lift was '*much cheaper than working a chain of locks*'. At the meeting on 21 March 1876, Sandeman said that, during the course of the construction process:

'he was much impressed with the readiness and confidence with which expedients were devised for overcoming the difficulties that necessarily presented themselves in such novel work'

and

'from an intimate acquaintance with the construction and working of this ingenious canal lift, he was pleased to be able to confirm what the Author [of the paper being discussed, i.e. Duer] *had stated as to its complete practical success.'*

Even with the cost over-run, which often happens on engineering projects even today, the completed structure was hailed by the engineering fraternity as a great success. The 1875 structure was elegant in its simplicity and was much admired at the time and thereafter.

★★★★★★★★★★

It is perhaps now appropriate to give a detailed description, with the assistance of diagrams, of the various elements of the original 1875 hydraulic lift structure. Regrettably, there is a risk that, for many readers, it will end up being rather a tedious list of facts and figures, a situation I will do my best to avoid. Any book about the Anderton Boat Lift would, however, be incomplete without it. Additionally, description of the lift is made rather

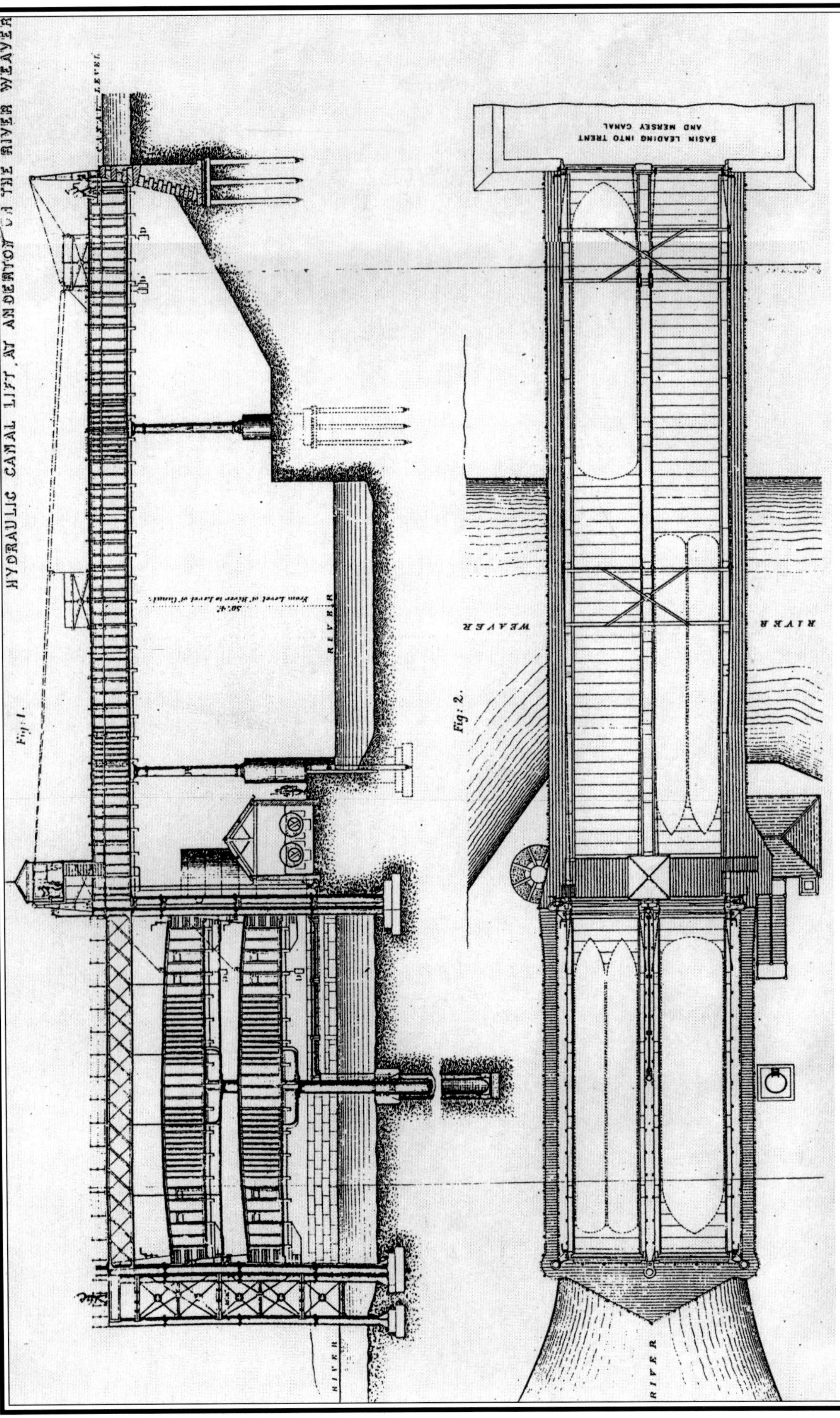

Plan and section of the hydraulic boat lift as designed by Edwin Clark and completed in 1875. Note the relatively small control cabin above the river end of the aqueduct, which was replaced with a larger cabin circa 1902. Courtesy The Institution of Civil Engineers

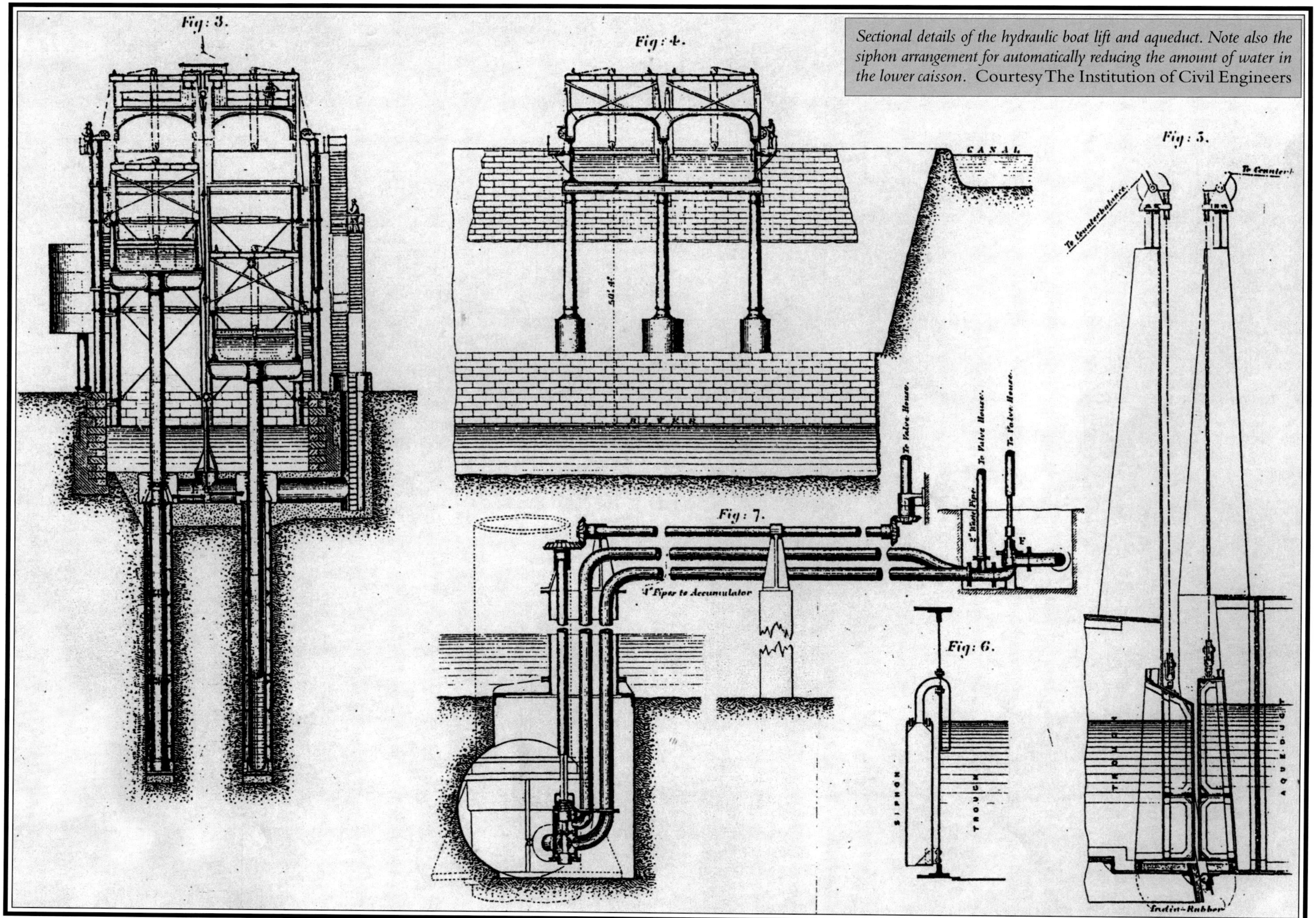

Sectional details of the hydraulic boat lift and aqueduct. Note also the siphon arrangement for automatically reducing the amount of water in the lower caisson. Courtesy The Institution of Civil Engineers

PLAN SHOWING ARRANGEMENT OF WHARVES AND WAREHOUSES AT ANDERTON C1890

Note Mr Thompson's warehouse next to the aqueduct; brine seeping from salt stored in the warehouse caused the bank near the lift to move in 1896. New piling was provided and the salt chutes realigned. Courtesy Cheshire Records Office, Chester

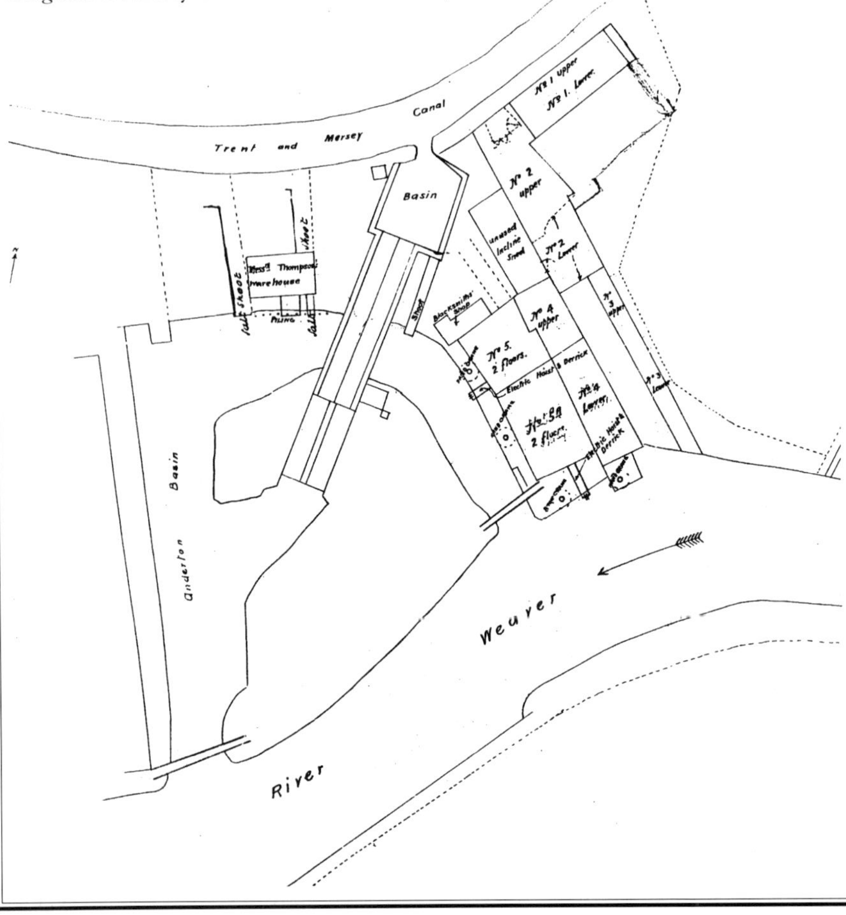

filled chamber at river level, formed from sandstone retaining walls on three sides, with the fourth side, facing south, open to provide the link with the Anderton Basin and the river. The lift chamber was sufficiently large to accommodate two wrought iron caissons, or boat tanks, side by side, each 75 feet long by 15 feet 6 inches wide. The caissons could, in turn, either accommodate two 72 foot narrowboats, with a beam or width of 6 feet 6 inches and a laden depth below water level, or draught, of 4 feet, or a barge with a beam of 13 feet or so. At each end of each caisson was a manually-operated, vertical-lift gate, which was raised to allow boats in or out and lowered to create a watertight seal, to enable the caisson and its contents to be raised or lowered. Each caisson, with water and with or without boats, had a nominal weight of 250 tons.

The caissons were made from wrought iron and had floors strengthened by gusset plates radiating from the centre. Each caisson was supported centrally from below by a single massive cast iron press, or hydraulic ram, the body of which was sunk vertically into the ground to a depth of nominally 50 feet below the bed of the lift chamber. The piston, or moving part, of each cylinder, consisted effectively of a tube made up from three bolted together lengths, with an outside diameter of 3 feet and a wall thickness of $1^{3}/_{4}$ inches. The pistons passed vertically through stuffing-boxes in the bottom of the lift chamber and then through a wrought iron access tunnel, before entering the presses or tops of the ram cylinders. The access tunnel had a diameter of 4 feet 4 inches, which enabled man-access to the valve-box, the glands of the cylinders and the stuffing-boxes for adjustment and lubrication. The access tunnel, although being below the water-filled lift chamber, was kept dry by means of the sealing effect of the stuffing-boxes.

complicated by the fact that some elements, such as the hydraulic rams and accumulator, no longer exist having been superceded at the time of the conversion to electrical operation in 1908, whilst other elements of the original structure remain almost unchanged. It is therefore necessary, for the sake of grammatical correctness, to slip between the past and present tenses during the course of describing the lift.

A small island, created by the piecemeal development of the Anderton Basin, as described in Chapter 3, was chosen by Leader Williams and the Weaver Trustees as the site of the new boat lift. This island, and hence the boat lift, was approximately midway between the Trent & Mersey Canal and the River Weaver, which at Anderton are horizontally just 400 feet or so apart, whilst, it should be remembered, the vertical distance between the two waterways is 50 feet 4 inches. Access to the boat lift from the river was relatively straightforward by means of the Anderton Basin, whereas access between the lift and the canal was more problematical, and necessitated the construction of the aqueduct.

The actual boat lift consisted of a rectangular, water-

The ram cylinders were installed within vertical shafts below the access tunnel, which were each formed by eight lengths of cast iron liners with a diameter of 5 feet 6 inches. Similar to the pistons, the cylinders were made up of three cast iron tubes with flanged, bolted connections. They extended virtually to the full depth of the shafts and were fitted into cast iron bed-plates, and held in place laterally by concrete placed between the cylinders and the shafts. The cylinders were interconnected hydraulically within the access tunnel by 5 inch diameter wrought iron pipework, which included isolating valves and connections to a steam-operated accumulator, or pressure vessel. The glands and stuffing-boxes which formed the joints in the hydraulic pistons were:

'packed with best quality spun yarn, which is plaited by the attendant as required, and softened with fine Russian or American tallow.'

The hydraulic rams operated at a pressure of 550 to 670 lbs. per square inch.

The vertical movement of the caissons was guided at each corner by hollow cast iron columns, three at each end of the lift basin, with an additional column at the river end to provide lateral stability and to support a high level working platform. The outside corners of each caisson incorporated cast iron guide blocks which coincided with angle guides on the inside faces of the guide columns. These were founded below the bottom of the lift chamber, on concrete and brickwork footings, and were built into the chamber's masonry walls. The triangular, high level working platform, at the south end of the lift structure, was linked to the aqueduct by means of wrought iron and timber walkways, which ran between the columns along both sides of the lift. The platform and walkways were strengthened by wrought iron lattice girders. A multi-stage wrought iron stairway at the north east corner of the structure enabled the lift operators to move easily between the elevated parts of the lift and the island.

The accumulator, a means of storing hydraulic energy without the need for water towers, was installed at the north west corner of the lift structure. It was kept continually primed by a 10 horse power, coal-fired, steam engine, which was housed in a brick building beneath the aqueduct. A tall, square, brick chimney was built to the east of the aqueduct, to dispel the smoke and steam from the engine.

At the river entrance to the lift are two small brick buildings, one on either bank, which were provided for the collection of tolls from boats using the lift.

The aqueduct, which generally is still in its original form, consists of a wrought iron water-filled trough, with a dividing central wall to form two separate channels, which coincide with the caissons on the lift structure. The aqueduct is supported in three spans by hollow cast iron columns. The length of the spans from the canal end are 57 feet 6 inches, 75 feet and 30 feet respectively, giving an overall length of 162 feet 6 inches between the boat lift and the canal basin. The overall width of the aqueduct is 34 feet 6 inches, with a depth of 8 feet 6 inches, whilst the normal depth of water in the aqueduct is 5 feet 3 inches. It has a total weight, including the water, of 1,050 tons or 6.5 tons per linear foot.

Each end of the aqueduct could be closed off by means of vertical-lift, wrought iron gates, which were counterweighted to facilitate manual operation. In addition, the aqueduct was provided with bracing against the outward pressure of the water by means of external gussets and overhead wrought iron gantries. Watertight joints were achieved between each caisson and the aqueduct by means of matching bevelled faces with rubber seals, which were compressed as the caisson approached its fully raised position.

A gantry over the aqueduct adjacent to the lift structure once supported a relatively small, timber weather-boarded valve hut, from which the lift operator opened and closed the various valves necessary to raise and lower the caissons. From the vantage point of the valve hut, the operator would have had unobstructed views to the north, overlooking the aqueduct and the basin on the Trent & Mersey Canal, and to the south overlooking the lift, the Anderton Basin and the River Weaver.

Similar to the boat lift, the aqueduct originally had wrought iron and timber walkways along each of its sides, to provide a link between the lift and the canal basin.

The final element in the structure was the so-called '*side dock or enlarged cut*', which consisted of a rectangular basin with sandstone walling and puddle clay lining, and which connected the aqueduct to the canal. The basin had a narrow entrance from the canal to facilitate bridging by the towpath and temporary stoplogging. The canal basin was sufficiently large to permit several narrowboats to moor up, in readiness for moving onto the aqueduct. A watertight seal between the sandstone abutment of the canal basin and the aqueduct trough was achieved by building a cast iron bed-plate, onto which the aqueduct rests, into the masonry of the canal basin and the use of timber wedges and cement slurry as caulking.

★ ★ ★ ★ ★ ★ ★ ★ ★ ★

The main part of the operation of the lift was simplicity itself, making use of the paradox of the '*equilibrium of liquids*', as first expounded by Blaise Pascal over a century earlier. Perhaps the clearest way to describe the lift operation is to use the analogy of old fashioned kitchen weighing scales, where each pan represents one of the caissons containing water and boats. Initially, each would be full of water and, whether containing boats or not, would be of identical weight, due to the boats displacing their own weight of water from the caissons. Under such circumstances, the system would be in equilibrium. If water was then taken out of one of the caissons, sufficient to overcome any frictional forces in the system, the heavier caisson would descend with the consequent compression of its ram, whilst the lighter caisson would be forced upwards with its ram being extended. During this procedure, the volume of water in the enclosed system of the hydraulic rams and interconnecting pipework would remain unchanged.

In practice therefore, consider the situation when one caisson is at the top of its travel, at the aqueduct level, with 4 feet 6 inches depth of water, and the other is at river level, with 5 feet of water. The lower caisson would then have 6 inches (15 tons) of water siphoned off automatically, so that it contained 4 feet 6 inches depth of water, whilst the upper caisson would be positioned so as to enable an extra 6 inches of water to be drawn from the aqueduct. At this stage the isolating valve on the pipework between the two rams would be closed, thereby preventing any movement of the rams. In this situation the gates at the appropriate end of each caisson and on the aqueduct could be raised, thereby permitting boats to enter or exit the

Many of the photographs of the hydraulic boat lift are undated and contain little information which can help the researcher be more precise but this view is certainly quite a bit later than the 1880 view on page 10 and is probably circa 1900 – a further salt platform and chute has been built on the right, just behind the distinctive brick chimney provided for the steam engine which powered the accumulator. Note also the lack of cover over the boilers and the pair of guillotine gates stacked against the shed; were they new replacements, worn-out examples or just removed during maintenance?

Courtesy The Boat Museum, Ellesmere Port

Emmerson, Murgatroyd & Co's cast iron builder's plate of 1875, still in place on the lift today. Photo Lewin, Fryer & Partners

caissons. Once loaded, the gates would be lowered into their closed positions, and the small chamber between the upper caisson gate and aqueduct gate drained of water. Then, by opening the isolating valve between the rams, the upper, heavier, caisson would descend, whilst the lower caisson would rise, the speed of movement being controlled by the rate at which water could flow through the 5 inch diameter pipework between the rams. The movement of the caissons would continue until such time as the descending one came into contact with the water in the lift chamber. At this stage, the caissons would slow down and eventually come to a stop, when the upthrust caused by the displacement of water on the one descending, was sufficient to counter the differential in weight between the two. The isolating valve would then be closed and sufficient water run off from the appropriate ram, so as to permit the descending caisson to drop under its own weight, until the water level in it was 6 inches higher than that in the river basin. The differential would then be siphoned off, as before, and the caisson river-gate raised to allow the boats to move out into the Anderton Basin.

The ascending caisson, which had come to a halt marginally below the required position, would then be raised by connecting its ram to the accumulator, until such time as the caisson floor was 5 feet below the aqueduct water level. The operation of small penstocks would then permit canal water to fill the gap between the aqueduct and caisson gates, and to equalise the levels between the aqueduct and caisson. The aqueduct gate and the caisson gate could then be opened to enable boats to move out onto the aqueduct. The procedure would then be repeated.

The whole operation of moving boats up from one waterway to the other could be completed in less than 10 minutes, with minimal power usage and very little loss of water from the canal. Furthermore, the actual lifting/lowering of the caissons took just 3 minutes or so. This compares very favourably with the time of one hour or so which a single boat would take to pass through a flight of five locks. In addition, a boat passing down such a flight would have resulted in the transfer of more than a lock-full of water from the upper waterway to the lower (one lock-full equates to about 25,000 gallons or 112 tons of water).

Occasionally, under certain circumstances, it was necessary to take one of the caissons out of service, say for maintenance or repair. At such times, the remaining caisson could be operated separately by means of the accumulator, albeit with an increased energy input and with a significantly longer lift time of 30 minutes or so to raise the caisson from the river level to the canal level.

The constructional materials used in the 1875 structure were almost certainly the best available at the time. The sandstone blocks used in the canal basin, the river basin and the lift chamber would have been readily available from local sources. Cast iron, which, by that time was being produced to high standards of quality, was used for the main compressive elements above ground, including the columns to both the lift and the aqueduct. Both stone and cast iron, however, are very brittle in nature and consequently are unsuitable for use in tension. For the aqueduct and caissons, therefore, Clark turned to the use of wrought iron, which has a lower carbon content than cast iron and is hence more malleable. As such it is more capable of withstanding tensile forces. However, one of the attributes of both stone and cast iron is their ability to resist corrosion, whether from oxidisation or chemical attack. Significantly, wrought iron is more vulnerable to these corrosive influences.

★ ★ ★ ★ ★ ★ ★ ★ ★ ★

In concluding this Chapter on the construction of the hydraulic lift, I refer again to the meeting at the Institution of Civil Engineers on 21 March 1876, when Clark described the operation of the boat lift as follows:

'the novelty consisted in lifting a piece of the canal itself, and not simply a vessel on a pontoon. Practically, the canal was continued over the river until it became vertical above the spot to which the barge was lowered, and a piece of the canal descended with the barge in it; similarly that same piece rose again.'

INFORMATION SOURCES FOR CHAPTER 4

Minutes of the Proceedings of the Weaver Navigation Trustees - 1908 to 1948. *British Waterways Regional Office, Northwich.*

The Anderton Boat Lift - An Examination of its Impact on Local Trade and Population, and of the Extent to which Trading Figures and Working Conditions Reflect National Economic Trends, *dissertation by Sharon Rosalind Murray, Chester College of Higher Education, 1990.*

Hydraulic Canal Lift at Anderton on the River Weaver. *Paper by Sidengham Duer. ICE Proceedings, Vol. 45.*

Narrowboats heading to Anderton, probably on runs for Potteries materials, photographed preparing to leave an overnight mooring on the Trent & Mersey in the mid 1920s. A couple of the boats are sheeted up, indicating that they may well be laden with cargoes of pottery. Note the horses feeding from their nosebags in the right background. There is no definite location with this photograph but it would certainly have been near a pub! Courtesy The Boat Museum, Ellesmere Port

CHAPTER 5
OPERATION OF THE HYDRAULIC LIFT (PART 1)

'Since the apparatus was opened by the Trustees, in the beginning of July 1875, no alterations or additions have been required; and great satisfaction is expressed in the way in which the lift works, and the ease with which it is manipulated.'
Extract from paper entitled Hydraulic Canal Lift At Anderton, on the River Weaver *by Sidengham Duer B.Sc., Assoc. Inst. C.E., 21 March 1876.*

Following its formal opening to traffic on 26 July 1875, as testified by the above quotation the hydraulic lift was, in its first few years at least, very successful and was recognised by many as an engineering masterpiece. It was acclaimed as a success even during the Annual Survey of the Weaver Navigation in July 1875, when the Trustees took the opportunity to inspect the recently completed boat lift. An article in *The Northwich and Knutsford Guardian* on 21 July 1875 gives an account of the survey and contains the following statements regarding the lift:

'The Lift presents a wondrous specimen of engineering skill, and is, we are informed, the only one of the kind in the world. It was designed by Mr. Edwin Clark and Mr. Sidengham Duer, of London and the contractors for the extensive work were Mess'rs. Emmerson, Murgatroyd and Co, of Stockport.'

Interestingly, the design of the lift is ascribed to both Clark and Duer. The article goes on to describe the operation of the lift in the following manner:

'For the purposes of thoroughly and practically explaining to the Trustees the working of the great lift, two laden canal boats were floated into each caisson, and the operations of ascending and descending were most successfully performed; the period during which a caisson was actually in motion at one time being about three minutes. The working of the whole of the huge machinery, under the direction of Mr. Sands was simply perfect and elicited many expressions of approval and admiration from the visitors.'

Nine years later, then Engineer to the Trust Lionel B. Wells still, seemingly, only had words of praise for the boat lift:

'It is worked with great economy. It does all the work brought to it day by day without being "sick or sorry," and since it was opened the lift itself has only caused one slight stoppage of traffic .... The works generally are of quite a unique character, and have been visited by distinguished foreigners, Colonial governors, engineers, and scientific men from all parts of the world.'

This effusive statement appeared in a paper entitled *Visit to Northwich, the Hydraulic River and Canal Boat Lift at Anderton and the New Flood Sluices and Locks at Dutton* which was presented to the Manchester Association of Employers, Foremen, and Draughtsmen on 9 August 1884.

The hydraulic lift, as designed by Edwin Clark, perhaps with assistance from Sidengham Duer, was elegant in its simplicity. Although the weight of each caisson when full of water was of the order of 252 tons, it was supported from below by a single, relatively slender, ram which had a fully extended length of over 50 feet. The guide columns were Romanesque, whilst the inter-linking iron latticework was attractively lightweight. The mode of operation made use of basic hydraulic principles and was beautifully simple, as all the best inventions tend to be.

Only one, relatively obscure, aspect of the design might be said to have impaired the workings of the 1875 lift and that was the adoption of a wet or flooded lift chamber as opposed to a dry chamber. This resulted in the need of a steam-driven accumulator to achieve the last foot or two of movement of the rising caisson. In the subsequent designs for the boat lifts in Belgium and France, this perceived imperfection was corrected by the incorporation of dry wells. Despite this relatively minor imperfection, the original hydraulic lift at Anderton can surely only be described as a masterpiece of Victorian engineering.

★★★★★★★★★★

Rather surprisingly, the day-to-day operation of the hydraulic lift was managed on behalf of the Trust by a relatively small team of men. In his paper to the Institution of Civil Engineers on 21 March 1876, Duer stated that:

'The staff required consists only of one man in the valve house, one engine driver, and three labourers for attending to the troughs and taking boats into and out of the lift. The total working expenses, including stores and a clerk to check the tonnage, are under £10 per week.'

The duties of the labourers would have included the manual operation of the various gates and penstocks on

As mentioned in the text below, one of the duties of the liftmen, outside of their work in operating the boat lift, was to help on ice-breaking duties around the Anderton Basin in wintertime. Canals freezing over was a drawback to their usefulness which all companies fought to overcome, although only with limited success, and it remained a problem right to the end of the carrying days, particularly on the narrow canals. This mid 1960s view shows a British Waterways boat trapped in ice on the Trent & Mersey, although the crew look fairly resigned to their predicament – it appears to be a maintenance craft anyway.
Courtesy The Boat Museum, Ellesmere Port

A pair of 'joshers', working narrowboats belonging to Fellows, Morton & Clayton, well known Midland's-based canal carriers, about to leave the eastern caisson, probably in the 1930s. The nickname was in affectionate remembrance of Joshua Fellows, son of company founder James and the man responsible for building the business into one of the largest carriers on the canals. This interior view, from the boatman's perspective, would have changed little over the years since the lift was first opened, despite its 1908 rebuilding. It can be seen that the butty boat on the left is named Urmston, *which thus identifies the motor boat as* Ferret; *these craft were built as pairs – these two in 1926 – and stayed together.*
Courtesy The Boat Museum, Ellesmere Port

both the lift and the aqueduct, and directing the movement of barges and narrow boats. When not involved on the actual operation of the lift, the men would have been employed on other activities on the structure and elsewhere around the Anderton Basin, including routine maintenance and repairs, loading and unloading prior to transshipment of goods, and ice-breaking during cold weather. For the first twenty years or so, the hours of working on the lift were from 6 a.m. to 10 p.m. from Monday to Friday, with the men working in shifts. In the early years there was no weekend working.

During the meeting on 21 March 1876, John Watt Sandeman advised of the working costs of the lift during the first year of operation and compared them with the maximum potential income, as follows:

'With the lift in full operation the working expenses amounted to about £15, to which should be added £93, or 10 per cent on the prime cost, being 5 per cent for interest and 5 per cent for depreciation, making a total of £108 per week . . . . The lift was capable of transferring sixteen barges, (narrow boats), per hour, eight up and eight down, and, proceeding upon the basis assumed,

this would give four hundred and eighty light and four hundred and eighty laden barges transferred per week; the laden barges at present averaged about 25 tons burden each, or 12,000 tons of goods transferred per week, which, divided into the average working cost per week, gave 2.16d. per ton. The parliamentary tolls were:

	s.	*d.*
Per ton for all goods	*0*	*1*
For each laden barge	*1*	*0*
For each light barge	*2*	*6*

Tolls for one week's traffic would therefore be:

	£
12,000 tons at 1d.	*50*
480 laden barges at 1s.	*24*
480 light barges at 2s. 6d.	*60*
	134

If, therefore, the lift were in full operation, the tolls would more than cover the working expenses, interest on the first cost, and charge for renewal. The lift was likely to prove much more remunerative by attracting trade to the navigation which it did not possess before; it had already done so to some extent, and this trade paid toll for the navigation independently of the tolls charged for the use of the lift.'

As might be expected with any new transport link, especially one on which the users were required to pay tolls, use of the lift in the first few years of operation was relatively low, with the bulk of the cargo moved by it being conveyed down from the Trent & Mersey Canal to the River Weaver. The greater part of these early cargoes would have been salt on its way to Liverpool via the Weston Docks. However, as local industry and traders adapted to the improved transport routes offered by the lift, tonnages increased year on year with only the occasional downturn, generally related to a specific circumstance. Raw materials bound for the Potteries, such as clay, flint and coal, and finished wares from that region, gradually grew in importance with respect to the commercial viability of the lift.

The tables below illustrates the viability of the lift by means of the recorded tonnages of goods passing through it for the 33 years of operation prior to its conversion from hydraulic control to electrical control in 1908.

In general, the annual tonnages given in the table relate to the 12 month accounting periods from 1 April to 31 March. However, the tonnage for 1875 relates to the nine month period from the time the lift opened in July 1875 to 31 March 1876.

Before proceeding with a detailed description of the operation of the hydraulic lift, it is worth making a few specific points with reference to the recorded annual tonnages:

1876 - Based on the maximum capacity of the lift, as suggested by Duer, of 480 laden narrow boats a week and 25 tons of cargo per narrow boat, in its first full year of operation the lift was working at approximately 5% of its maximum capacity. With the initial charge rates of 1d. per ton of cargo, and 1s. for each laden boat and 2s. 6d. for each empty boat, the annual income in the first full year of operation would have been approximately £250. In comparison, the working costs on the lift would have been

RECORDED TONNAGES OF CARGO PASSING THROUGH THE ANDERTON BOAT LIFT 1875 - 1907

Year	Tonnage	Year	Tonnage	Year	Tonnage	Year	Tonnage	Year	Tonnage
1875	17,028	1882	22,066	1889	66,989	1896	154,849	1903	191,317
1876	31,294	1883	48,339	1890	59,688	1897	169,449	1904	191,886
1877	30,886	1884	49,722	1891	76,600	1898	186,035	1905	191,886
1878	37,363	1885	67,783	1892	87,763	1899	186,571	1906	192,181
1879	37,874	1886	75,942	1893	84,434	1900	181,276	1907	166,043
1880	40,097	1887	78,611	1894	68,920	1901	177,394		
1881	44,847	1888	75,375	1895	129,187	1902	177,931		

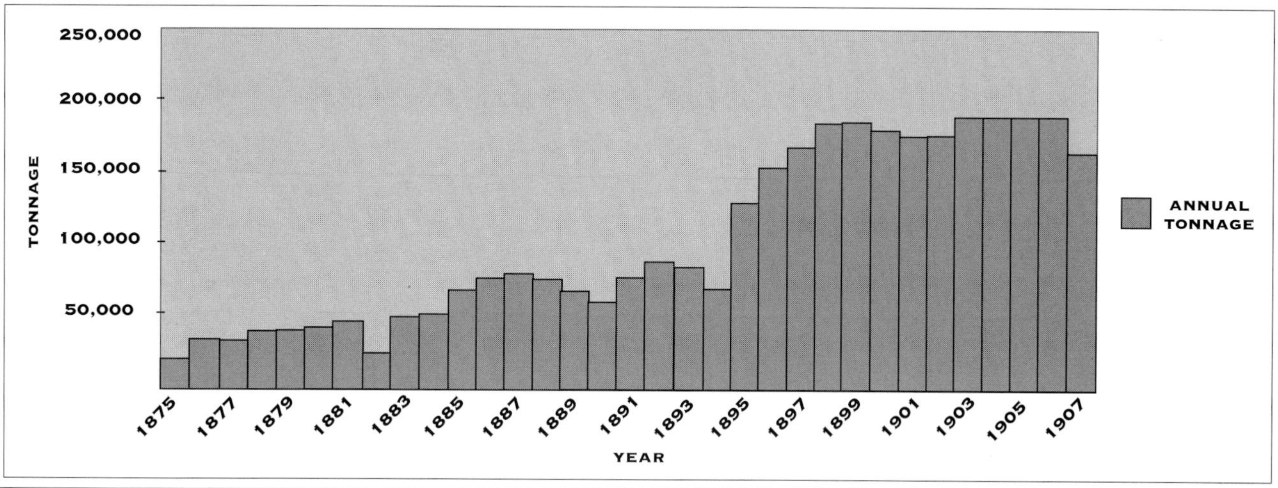

Another undated photograph of the Anderton Basin, with a laden narrowboat, probably carrying flints for the Potteries, about to enter the boat lift. On the left, at least two Weaver flats can be made out loading salt beneath the chutes. Note the slender nature of the hydraulic lift – in contrast to the converted electrical lift we know so well today – which looks almost in keeping with the spidery construction of the salt chutes. The glassy appearance of the water, caused by a long camera exposure, would suggest this is a circa 1890 view. Note the small control cabin atop the structure; later photographs show this was replaced by a larger cabin when Saner carried out the conversion of the lift gates to electrical operation and other works, circa 1902. Courtesy The Boat Museum, Ellesmere Port

close to £750. But to tempt more traffic through the new lift, the Trustees halved the charge on boats using the lift when the same carrier owned the two boats in one caisson.

1877 to 1881 - The next five years of operation saw a steady growth in trade with annual tonnages increasing over that period by nominally 50%. This was presumably due to the gradual change in transport patterns as traders began to realise the benefits of the improved link between the canal and the river.

1875 to 1887 - The cargoes through the lift in those early years were principally white salt, rock salt and raw materials for the Potteries, i.e. flint and china clay.

1881 - The north west of England witnessed several very cold winters when traffic, firstly on the canal, then the river, and, as an inevitable consequence, on the lift as well, came to a halt due to heavy frosts and thick ice on the waterways. For example, in his Annual Report to the Trustees on 26 July 1881, Wells stated that:

'Intense cold in January, from 12th to 26th during which time the mean minimum temperature at Northwich was 10 degrees or 22 degrees below freezing closed the inland navigations of this country. Weaver Navigation closed from 21st to 28th and Trent and Mersey Canal from 15th to 5th February.'

1882 - A dramatic downturn in total tonnages occurred as a result of a failure on the lift which caused its closure for a period of nominally six months (see below).

1883 - The annual income to the Weaver Navigation from the lift had risen in 1883 to just under £300. Traffic volumes rapidly recovered from the downturn in the previous years.

1886 - Traffic volumes increased steadily after 1886 when the Weaver Trustees declared that the Anderton Lift was:

'free of Toll for all vessels whose cargoes pay the ordinary Weaver Dues, in the hope that this will bring more traffic onto the River.'

1888 - Another downturn in traffic when salt tonnages carried on the lift fell from 22,082 tons in 1887, to 13,451 tons in 1888, and 5,449 tons in 1889. This was primarily as a consequence of Brunner Mond gaining parliamentary approval, despite opposition from the Weaver Trustees, to lay brine pipes under the river from Witton to their works at Winnington, which faced the lift from the far side of the river.

1880 to 1899 - The volume of raw materials passing through the lift bound for the Potteries continued to climb steadily during the 1880s and 1890s but the volume of salt being carried made a partial recovery only after 1888.

1890 - The Engineer's Report to the Trustees stated that:

'In the months of December and January the temperature fell lower than it has done for many years and the cold was very intense. The average temperature at Northwich for the two months, as taken by a shaded thermometer, was not quite one degree above freezing point. There were 32 consecutive days on which there was frost but the lowest temperature recorded was on December 22nd when the shaded thermometer registered 27 degrees below freezing point . . . . Anderton Lift was, however, not worked from December 20th to January 26th, owing to the Trent and Mersey Canal being completely frozen over.'

1894 - Tolls were raised in from 8d. to 1s. per ton.

1895 - Total tonnages of cargoes on the lift saw an upturn in 1895, probably as a result of the passing of the Canal Tolls & Charges Act of 1894, which encouraged pottery manufacturers to ship their finished wares to Liverpool via the River Weaver from Anderton, as opposed to using the Trent & Mersey and Bridgewater Canals to Runcorn.

1895 to 1906 - The volume of goods carried increased erratically during this period, reaching a peak of 192,181 tons in 1906. Thereafter, severe corrosion on the hydraulic rams resulted in increasingly frequent stoppages for repairs and, consequently, disruption to traffic.

1875 to 1908 - The total volume of traffic passed through the hydraulic lift in its lifetime was nominally 3.4 million tons, with an average yearly tonnage of approximately 103,000 tons

★ ★ ★ ★ ★ ★ ★ ★ ★ ★

Twelve years after the opening of the lift, in a paper entitled *Notes on The River Weaver Trust*, read before the Northwich Debating Society in 1887, Thomas Ward J.P. stated that:

'Undoubtedly the Anderton hydraulic lift is a splendid work, and answers its purpose admirably. It has been visited by engineers from all parts of the world, and is being imitated in various places . . . . If its usefulness were to be measured by the tolls it produced it must be considered a failure; but if it be looked upon as part of the work necessary for the proper carrying on of the trade on the Weaver, like a lock or a dock, it must be considered a great success.'

Seemingly, the lift worked faultlessly for the first five years or so, although it is hard to imagine that the early years of such a novel structure could have been totally trouble-free. The first recorded problem, however, occurred in February 1881, when some debris became jammed beneath one of the caissons while it was being lowered, with the result that the cover to the access tunnel was broken. Soon after, in July 1881, traffic in the lift was interrupted for seven days due to a fracture of the accumulator gland; (i.e. the seal between the moving ram and the fixed cylinder).

By that time, the position of Engineer to the Trust had again changed hands. John Watt Sandeman, who had replaced Edward Leader Williams as Engineer in October 1872, was himself replaced nearly five years later. On 5 March 1877, Lionel B. Wells was appointed as the new Engineer to the Trust on an annual salary of £800. Sandeman's departure may have been connected with his involvement in the design of a new dredger which, the year before, apparently keeled over and sank whilst undergoing trials at Weston Point. A short time prior to his departure, Sandeman was formally censured by the Weaver Trustees for his part in the demise of the dredger.

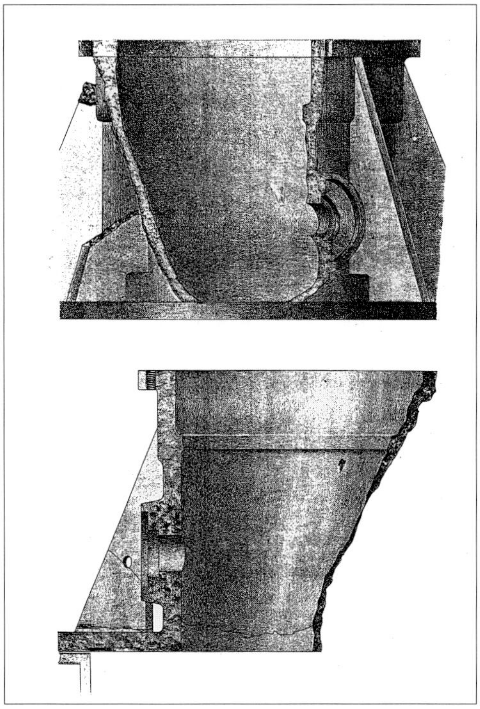

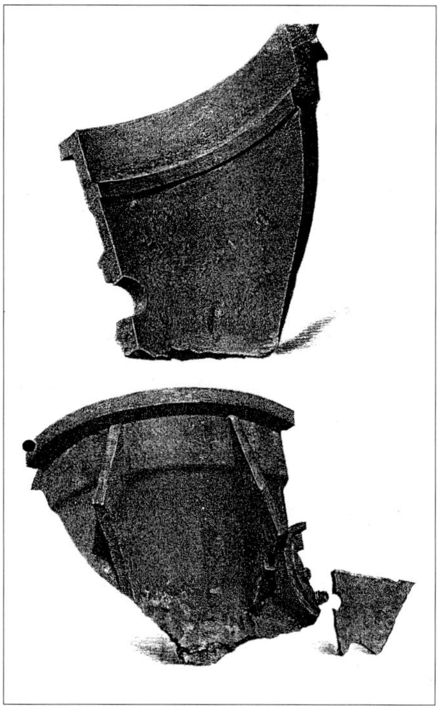

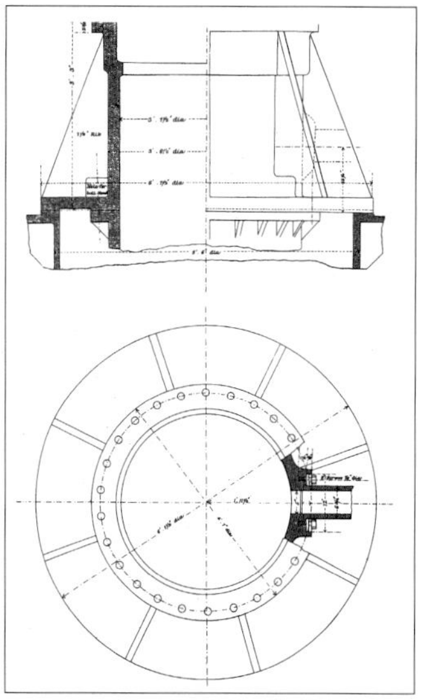

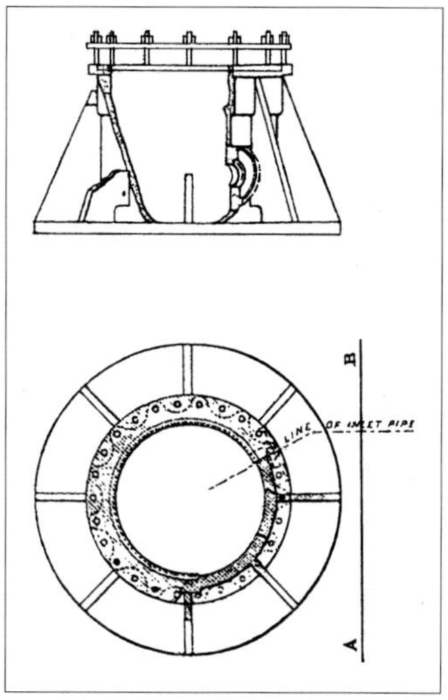

Above: *Detailed drawings of the fractured pieces of the ram presses after failure in April 1882. Taken from a hand-written report by Edwin Clark dated May 1882.*
Courtesy The Institution of Civil Engineers, London

Right top and middle: *Engineering drawings of the ram presses before and after failure in April 1882.* Courtesy The Institution of Civil Engineers, London

Below: *A diagram of the interconnecting pipework of the redesigned presses which was included in the minutes of a meeting on* River and Canal Works, *as contained in the* Proceedings of the Institution of Civil Engineers, Vol. 96: 1889. Courtesy The Institution of Civil Engineers, London

Figs. 1.

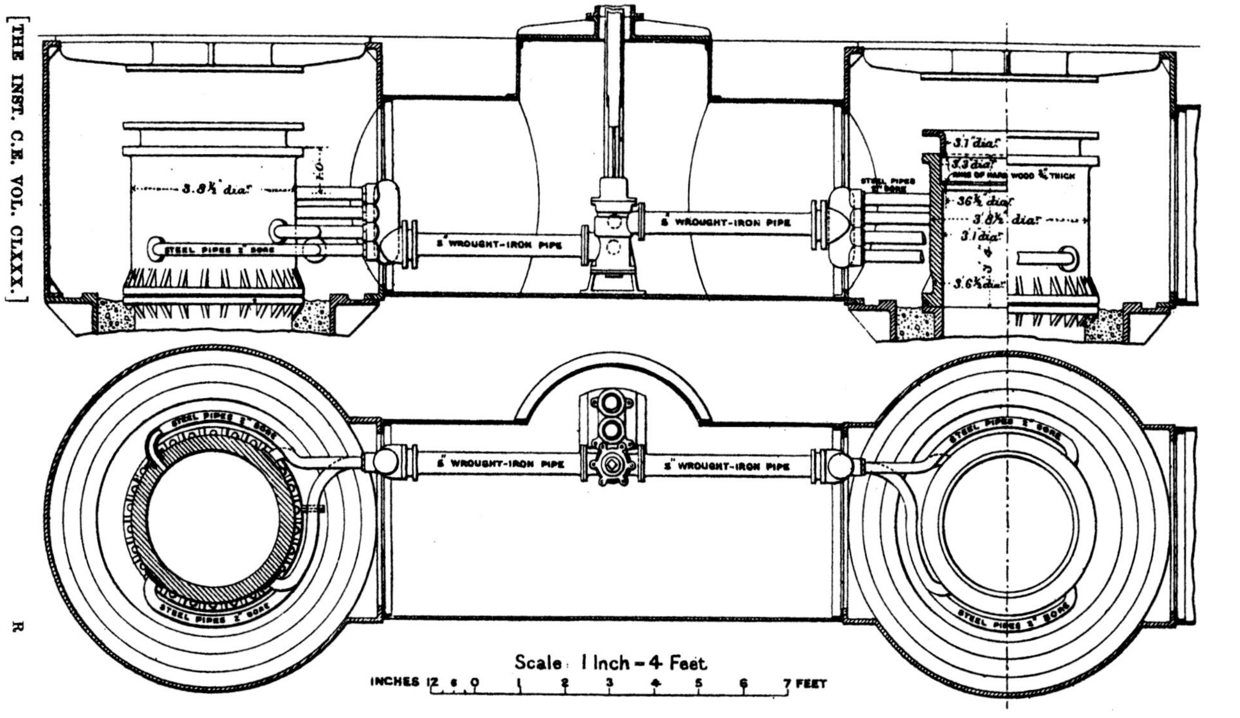

Notwithstanding the above, Sandeman's contribution to the Anderton Boat Lift is very much understated in available historical literature and, as such, he is very much the unsung hero. Edward Leader Williams, Edwin Clark and J. A. Saner are, quite rightly, names that immediately come to mind in connection with the boat lift. However, it should not be forgotten that Sandeman took over as Engineer from Leader Williams during the early stages of the lift's detailed design, with the latter being retained by the Trust in a consultative capacity. During his tenure, Sandeman presumably would have had a supervisory role in the design of the lift, by Edwin Clark of Messrs. Clark, Standfield & Clark, and was personally responsible for important changes to the construction of the foundations. Similarly, Sandeman must have played a significant part, alongside Leader Williams and Duer, also of Clark, Standfield & Clark, in the construction process.

Another major incident at the lift occurred in April 1882. Strangely enough, it was predicted, almost to the last detail six years earlier. At a the meeting of the Institution of Civil Engineers in London on 21 March 1876, a Mr. Bramwell is recorded as having expressed his concern regarding a perceived weakness in the design of the, then new, hydraulic lift:

'He thought that the incline had one advantage - that if anything gave way, the ratchets and palls provided a check; while if anything gave way in this lift, the whole affair would come down with a hideous crash. He could not help thinking that the press cylinders should have been of wrought iron (as opposed to cast iron). It was true the pressure per square inch was not great; but after all, cast iron was a treacherous material, and he should have liked to see the external cylinders made of wrought iron; or if cast iron were retained, it should be hooped. As a measure of precaution it might be well to have two other rams, one at each end, not to aid in raising the load, but simply to draw water into their cylinders as they rose, which water on their descent flow out through small apertures, sufficiently large to admit of the free exit of the water when the load descended at a proper pace, but so small as to check the outflow demanded by a too rapid descent, and in this way serious accident might be prevented.'

A very salutary warning but unfortunately too late to be acted upon even if the will had been there. Regrettably it all went horribly as predicted just a few years later. In his formal report to the Weaver Trustees on 21 July 1882, Wells, perhaps not surprisingly, understates the severity of the incident:

'On the 18 of April [1882], *an accident occurred through the bursting of the top length of press at Anderton. The caisson fell from the top into the water; fortunately no further damage was sustained beyond the bursting of the press.'*

However, he gave a fuller and more dramatic account of the event in the *Proceedings of the Institution of Civil Engineers, Volume 96, 1889*:

'From 1875 to 1882 the lift at Anderton worked without let or hindrance, when an accident of a very serious nature occurred, fortunately unattended with injury beyond damage to one of the presses. The up-river trough, or caisson, had been raised to the top, and the aqueduct gate lifted about 1 foot, preparatory to levelling up the water in the trough for passing a boat from the aqueduct, when a burst occurred, and the trough fell to the bottom. The water from the aqueduct rushed through the opening under the gate, into the descending trough, forcing the stern of the boat against the gate, and damaging the rudder, but beyond this, and some trifling damage to the upper works, there was no further injury to the boat or the lift. It was impossible to determine at what speed the trough reached the bottom; but although the head of the press was burst, the escape of water was regulated by the space between the ram and the press, the diameter of the former being 3 feet, of the latter 3 feet 1 inch; to this, and to the fact that the trough dropped on to a water-cushion, its freedom from damage was due. A workman was in the tunnel greasing the ram at the time, and, hearing a run of water, looked round and saw a stream spurting from the connecting-pipe and press; and he turned to go, when a burst took place; he was carried off his legs by the rush of water, but fortunately reached the bottom of the ladder, up which he scrambled. On pumping out the water from the tunnel, it was found that a side of the press had blown out and the pressure-pipe leading to it was cracked.'

It was certainly a very narrow escape for the un-named workman. The lift was out of action for almost six months while investigations, followed by experiments and, eventually, repairs were carried out. In the first instance, Wells sought help from those who had been involved in the original design process; namely Leader Williams, Duer and representatives from Emmerson Murgatroyd & Co. Their involvement is recorded in the minute book of the Weaver Trustees, as follows, from the *Summary of the Year's Business on 1st August 1882*:

'The Engineers and the Contractor employed in erecting the lift were called in and consulted as to the cause of the accident but they did not agree. We decided to test the ram of the other lift, and that burst just in the same place, with a pressure of 800 lbs. to the square inch, this proving that there was not sufficient strength. Another top length was then cast of greater strength which was proved to bear a pressure of 2,800 lbs. to the square inch. The remaining parts of the cylinders having also been proved, we decided to have two upper lengths of the cylinders re-cast by Mess'rs. Emmerson, Murgatroyd and Co. and refitted as early as possible.'

Rather belatedly perhaps, Edwin Clark, the original designer of the lift, who was then 68 years of age and in semi-retirement, journeyed to Anderton on 6 June to examine the fractured press and to offer expert advice. The impression gained is that Clark was delayed in attending due to his being on the Continent at the time of the incident, because, as well as being accompanied by Wells, three engineers from the Belgian Government were also in attendance during his visit. His partners, Latimer Clark (Edwin's younger brother) and Mr. Standfield, had already inspected the fractured press. By that time, it can be assumed that Messrs. Clark, Standfield & Clark were already working on the boat lifts in France and Belgium, for which the Anderton Boat Lift was the forerunner.

During the six-month closure period, a vertical hoist was constructed and temporary rails were laid up the steep slope between the river and the canal, primarily for transferring ascending cargoes in the absence of the lift. Presumably, during this period, an increased volume of salt was transferred down from the canal to the river by means of the chutes.

In the Engineer's Report to the Trustees on 30 July 1883, Wells recorded that new top lengths of the presses had been satisfactorily installed and tested during September 1882, with the lift re-opening to traffic on 5 October. He concluded by advising that, since then, the lift '*has continued to work in a satisfactory manner.*' The cost of the investigations and repair works to the Trustees amounted to £1,432 14s. 4d.

The official conclusion of the investigations, as recorded by Wells in the *Proceedings of the Institution of Civil Engineers* in 1889, was that the likely cause of failure of the first press and the subsequent failure of the second press during trials, was stress-induced cracking of the casing emanating from the connection with the 5 inch pipe linking the two presses. In the original casting, the presses had been strengthened around the connections by locally increasing the thickness of the cast iron casings. It was suggested that the casting of this thickened area was a difficult process and could well have led to local deficiencies in a critical area of the ironwork. Several other possible causes were mooted, such as settlement of the foundations of the lift due to underground salt abstraction and leverage on the press caused by sideways movement of the fully extended cylinders. These, however, were eventually discounted. In a letter to Edwin Clark on 27 May 1882, Wells states:

'I can find no trace of any settlement apparent, indeed the fact that the two presses have burst on their inner sides, that is one on the West and the other on the East side of the press, would seem to show that their failure was not due to any displacement of the press itself, the metal also seems to be good and as there is a relief valve at the top of the stroke, and the accumulator was at work I do not think that the extra pressure suggested by you can have been the cause of the first accident.'

Reference to other records dating from around the time of the failure suggest that in reporting these findings, Wells was being either cautious or diplomatic or both. Clark, however, having carried out his own investigations, reached conclusions that were not totally consistent with Wells' findings. In his report on the incident Clark records that:

'The castings which failed were defective and unsound and in one place had even been plugged with wrought iron by the Manufacturer to hide some of the defects. It had been at work many years before it broke with less than its ordinary normal strain, several boats having already been raised on the day on which the fracture occurred.'

Clark went on to explain that in addition to the defective ironwork, the presses had, over the seven years of operation, been displaced vertically downwards due to the compression of timber packing between the bottom of the rams and the base of the ram shafts. The amount of settlement was such that:

' . . . the flange (of the presses) had in both cases come into actual contact with the cylinders (lining the ram shafts) which are not truly vertical unfortunately bearing only at one spot thus throwing an unequal and oblique tensile strain on the large flange of the upper segment of the press which it was not calculated to support . . . . This tended to cause a distortion of the press especially at the point where the connecting pipe entered it, and where the fracture took place – After a careful consideration of all the circumstances I am clearly of the opinion that this was the determining cause of the fracture, and in the case of the first press this cause was heightened by the unsound character of the casting at the spot at which it broke – The second press which was subjected to similar strains supported a much higher pressure, probably owing to the sounder character of the iron.'

At the end of his report Clark states that Wells:

' . . . completely agrees with me as to the cause of the failure, although he attaches less importance to the quality of the metal which in the second press was very good.'

Clark's report, particularly his criticism of the original casting, resulted in a flurry of letters between the suppliers of the castings, Messrs. Emmerson, Murgatroyd & Co., and Messrs. Clark, Standfield & Clark. Seemingly Emmerson, Murgatroyd were only made aware of the situation when Edwin Clark's findings regarding the quality of the castings were published in a formal report. In a letter dated 11 July 1882 to Clark, Standfield & Clark at their offices in Westminster, J. L. Emmerson describes Edwin Clark's behaviour as being '*very unprofessional*'. Emmerson then goes on the offensive:

'I observe that Mr. Clark in this report forgets one not unimportant particular that on the 6th June he met Admiral Lord Clarence Paget and myself at the Crewe station and volunteered the information that he had just been to Anderton and had examined the castings, that one of them was faulty and of rather inferior metal and that this casting had had a piece of wrought iron let into it, but that the accident had nothing to do with the material but arose from faulty design and principally from the large diameter of the inlet hole, the other press he said was of excellent material.

On going to Northwich I found on the same day he had said the same thing to Mr. Lionel B. Wells who repeated it to Mr. Leader Williams, Mr. Duer and myself.

I therefore request that you will explain to me without delay how it is that the report differs so much from the conversation with Mr. Wells and myself.'

It is possible that Edwin Clark was again out of the country because in their response to Emmerson on 15 July, Clark, Standfield & Clark advised that his letter had been forwarded to Edwin Clark. Emmerson copied his letter to Wells who responded on 23 July in the following manner:

'I fully expected you would challenge Mr. Edwin Clark's report. He certainly did not attribute the accident to the quality of the casting when he was here, quite the reverse. I think the fact that your firm and only your firm was asked by the Trustees to supply castings for repairs of the lift is the best answer I can suggest.'

Having not received a reply from Edwin Clark, Emmerson wrote again to Clark, Standfield & Clark averring that he was '*determined not to let this matter rest in its present unsatisfactory state*'. It would appear that Emmerson did eventually receive a response from Edwin Clark but that it was clearly not to his satisfaction because on 3 August he wrote to the Trustees advising that:

'Mr. Edwin Clark's conduct in this matter has to speak mildly been most unprofessional, but when added to that, he makes gross mis-statements I cannot allow this matter to rest. If in a few days I receive no other reply, I shall forward the report and correspondence to the Secretary of the Institution of Civil Engineers requesting him to bring it before the Council.'

At this point, unfortunately, the trail disappears and there is no record of how the matter was resolved. The fact, though, that Clark's report and copies of the various letters are retained in the archive of the Institution suggests that Emmerson went through with his threat. It should be noted, however, that at the time of the incident Edwin Clark had been retired from active business for six years. It seems, therefore, that he was either working on a freelance basis in investigating the incident or, more probably, acting on a part-time consultancy basis for his old practice of Clark, Standfield & Clark.

Notwithstanding the above, the presses were re-designed to deal with the stress concentrations at the pipe connections and presumably steps were taken to prevent the flanges bearing on the casing of the rams' shafts. The use of cast iron was continued, but instead of having single 5 inch diameter connections, as on the original presses, the new presses were cast with four, thicker walled, 2 inch diameter connections. Other, less significant alterations, were made to the presses at the same time. The new, modified presses were, as expected, fabricated and installed by Emmerson, Murgatroyd & Co. It is interesting to note that the designs for the presses for the proposed lifts in France and Belgium were modified accordingly.

On 24 October 1882, Leader Williams, Wells and Duer sent a joint, handwritten, letter to the Trustees advising of the completion of the remedial works and making recommendations with regard to the future operation of the lift:

'To the Trustees of the Weaver Navigation

My Lords and Gentlemen,

We beg to inform you that the repairs to the Anderton Lift are now completed and that it was satisfactorily tested on the 2nd & 3rd inst. with a weight of water considerably in excess of its ordinary working load. All the operations required for the passage of boats were gone through in a satisfactory manner & we are of the opinion that the Lift may now resume its usual work.

We suggest that the following precautions should in the future be taken with regard to the Lift.

***1st** – That observations be made from time to time to see whether subsidences in the ground take place in the neighbourhood of the Lift.*

***2nd** – That the Lift be tested once every six months with water up to the top of the gates so as to be sure that the foundations are immoveable and that the pipe presses are in working order.*

***3rd** – That when the trough (caisson) is at the top of the Lift the gates at the canal end of the aqueduct should always be closed and that the gate next the boat whether it be in the Aqueduct or in the trough should not be lifted more than six inches until the other gate has been lifted sufficiently to allow the passage of the boat.*

We remain
My Lords and Gentlemen
Your Obedient Servants
E. Leader Williams
Lionel B. Wells
Sidengham Duer'

A year later, in his annual report to the Trust on 30 July 1884, Wells advised that '*this lift has been tested twice during the year and found in good working order*'. It would be reasonable to assume that the recommendations made to the Trustees were rigorously applied for the next three years whilst Wells remained as Engineer to the Trust. Whether they continued to be followed after his departure at the end of 1887 is not recorded but it is probable that, as is often the way, the precautionary measures were relaxed as the operators gained confidence in the sufficiency of the modified lift.

★ ★ ★ ★ ★ ★ ★ ★ ★ ★

Perhaps indicative of problems to come, a failure occurred on 23 March 1887, when the gland of the press at the top of one of the hydraulic rams burst and had to be temporarily repaired by means of a wrought iron plate. It was subsequently replaced four years later in June 1891 by a stronger casting, which was made up in three sections to facilitate installation. The gland of the other ram press was similarly replaced in April 1894.

★ ★ ★ ★ ★ ★ ★ ★ ★ ★

The hydraulic boat lift, since its opeing in 1875, had undoubtedly been a commercial success, drawing a significant amount of traffic, and hence revenue to the Trustees, onto the River Weaver from the Trent & Mersey Canal. However, whilst the early years of the operational life of the Anderton Boat Lift had been relatively trouble-free, the ingredients for a more insidious problem were present even at the time of the opening of the lift. Within ten years, a relatively short period in terms of the expected lifetime of an engineering structure, the cast iron rams, which raised and lowered the caissons, had become worn to such an extent as to give concern to Wells.

Two evocative 1930s studies of the western aqueduct; from the depth of freeboard showing, it is not the same vessel in each view. On the right of the bottom view, a boatwoman tends to a towing horse, whilst her young daughter watches the progress of the boat as it glides into the top basin with barely a ripple in the water.
Both courtesy The Boat Museum, Ellesmere Port

INFORMATION SOURCES FOR CHAPTER 5

Hydraulic Canal Lift At Anderton, on the River Weaver. *Paper by Sidengham Duer B.Sc., Assoc. Inst. C.E., 21 March 1876.*
Article in the Northwich & Knutsford Guardian, *21 July 1875.*
Visit to Northwich, the Hydraulic River and Canal Boat Lift at Anderton and the New Flood Sluices and Locks at Dutton. *Paper by Lionel B. Wells, 17 May 1884.*
Weaver Trustees Minutes – 1875 to 1895, *by Jack Horton, 1986.*
River Weaver Annual Accounts and Reports – 1881 to 1895. *Held at British Waterways' Regional Offices at Northwich.*
The Anderton Boat Lift - An Examination of its Impact on Local Trade and Population, and of the Extent to which Trading Figures and Working Conditions Reflect National Economic Trends, *dissertation by Sharon Rosalind Murray, 1990, courtesy Chester College of Higher Education.*
Notes on the Weaver Trust. *Two papers by Thos. Ward, J.P., 1887.*
The Weaver Navigation, *by Harry Wardale.*
Continental River and Canal Works. *Paper in the Proceedings of the Institution of Civil Engineers, Vol. 96, 1889.*
Correspondence between L. B. Wells, J. L. Emmerson and Clark, Standfield & Clark in 1882 and retained in the archive of the Institution of Civil Engineers in London.
Report into Incident at Anderton Boat Lift on 18th April 1882, *by Edwin Clark.*
Letter to the Trustees of the Weaver Navigation from E. Leader Williams, Lionel B. Wells and Sidengham Duer 24 October 1882.

Colonel J. A. Saner, Engineer to the Weaver Navigation Trust, 1887-1934. Edward Paget-Tomlinson collection

CHAPTER 6
OPERATION OF THE HYDRAULIC LIFT (PART 2)

'The grooving of the rams appears to have commenced in 1882, and it was all important to keep the stuffing box pressure tight, pieces of soft copper were dovetailed into the bad places, a method which was then being extensively used in the rams of the hydraulic graving dock at the Victoria Docks, London . . . .'
Extract from a paper entitled *Reconstruction of the Canal-Boat Lift on the River Weaver at Anderton* by J. A. Saner, 1910

In the above paper, Saner, the Engineer to the Trust at that time, gave a background to the corrosion problems which beset the Anderton Boat Lift only seven years after its construction and which eventually led to its conversion to electricity in 1908.

Fifteen years earlier, in a special report to the Finance and Works Committee of the Weaver Trustees on 30 January 1895, Saner speculated on the possible causes of the grooving as follows:

'A constant source of trouble with all hydraulic machinery is the longitudinal grooving action which takes place in rams, especially those which have been in use for some time, and it is this grooving which is now giving the trouble at Anderton . . . . Whether the cause of this action is due to softening of the metal through its constant exposure to water, to the acidity of the water, to the scouring of the 'packing' in the 'stuffing box' or to an unequal distribution of the constituent particles of the cast iron during the process of casting, has not yet been definitely ascertained.'

The main cause of the corrosion was evidently the use of the polluted canal water within the hydraulic operating system, combined with the water-filled basin at the bottom of the lift. Arguably, the quality of the cast iron in the rams was of secondary importance. It is now evident that the hydraulic lift was fifty years or so ahead of its time, the ram system requiring technology that was just not available at the end of the 19th century.

★ ★ ★ ★ ★ ★ ★ ★ ★ ★

The first reference to the grooving problem in the minutes of the Weaver Trustees was on 30 June 1885 when Wells stated in his annual report that, in the past year:

'The rams have been repaired, for which purpose the lift was stopped working for three weeks. Meanwhile arrangements were made for transhipping the pottery materials by steam.'

The reference to steam evidently relates to the use of steam driven cranes in the absence of the lift. As on previous occasions of reporting to the Trustees, Wells was economic with his use of words. Fortunately, he gives a fuller explanation in the *Proceedings of the Institution of Civil Engineers, Volume 96, 1889*, as follows:-

'In 1884 it was found that the rams between the top of the tunnel and the surface of the river-water, for a length of 8 feet, were becoming grooved or corded, to such an extent as to allow of the passage of an undue quantity of water, while the length was moving through the gland (the seal between the ram and the press).'

After these initial repairs, the lift seemingly functioned in a satisfactory manner for several years, although the problem clearly had not gone away. On 30 June 1887, Wells reported a minor failure that was almost certainly caused by the grooving of the rams:

'The gland of the up river press at the Anderton Lift gave way; this was repaired, and the lift restored to work in two days, meanwhile the traffic was passed by the other caisson.'

Unbeknown to Wells, in attempting a repair by means of introducing copper strips to the grooves, a further problem was inadvertently being introduced – that of bi-metallic or electrolytic corrosion between the cast iron and the adjacent copper, with the canal and river water acting as the electrolytic medium. The effect was to accelerate, rather than to retard the deterioration of the rams. Ten years later, Saner explained in his special report of 30 January 1895:

'The most frequently adopted method, (of addressing the problem), and one which is largely used at the Hydraulic Graving Dock, London, is to fill the grooves with nearly pure copper, small pieces of which are dovetailed into the cast iron by cutting away the soft portions, on identically the same principle as a dentist stops a tooth. There is no doubt that in most cases where good water is used, the life of a ram may be greatly prolonged in this way, but should the water be impure and acidulated with chemicals of different kinds, the contact of the copper with the cast iron sets up an electrolytic action, by which iron suffers decomposition more

rapidly than it would do if the copper were not present.'

The canal water, which at that time was being used within the ram system, was badly contaminated by a combination of the natural salt deposits, spillages during the transshipment of salt and pollution due to the growing industrialisation in the Anderton area. The resulting fluid was a cocktail of chemicals ideal for bi-metallic corrosion. Saner demonstrated by a series of experiments that, under the conditions at that time, the cast iron of the rams was being depleted by the electrolytic process.

However, in 1884 Wells was seemingly unaware of the detrimental effects of filling the grooves with copper and this method of repairing the rams was continued.

* * * * * * * * * *

On 25 December 1887, Wells was replaced as Engineer to the Trust by Saner, who until then had been Wells' assistant. Saner was appointed on an initial salary of £250 per annum. Wells took up his new post as Engineer to the Aire & Calder Navigation in Yorkshire but was retained by the Trustees for a period of one year to advise and assist Saner, on a salary of £200.

* * * * * * * * * *

Repairs to the rams by filling with copper strips continued during 1888 and 1889, with traffic through the lift being managed by single working of the caissons; a much slower and more expensive procedure which necessitated the use of the steam driven accumulator over the full height of the lift. Also around this time, experiments were carried out by Brunner, Mond & Co. with the use of tin, instead of copper, to fill the grooves in the rams. Tin, however, was found to be unsuitable because it was not sufficiently ductile, or soft, and could not be adequately worked into the grooves.

In the meantime, other, rather mundane but nevertheless essential, maintenance works were undertaken on the lift. Saner's end of year report to the Trustees on 30 June 1890 recorded that:

'While the Section of the Trent & Mersey Canal in connection with the Anderton Lift was stopped for repairs in August last, (1889), the portions of the Anderton Lift which are usually under water were thoroughly painted and repaired, the bearing piles for the caissons, (within the lift chamber), which had given way, being restored to their original position.'

In June 1892, yet more filling of the grooves in the rams was needed, together with the replacement of the gland for the up-river press. At the same time, alterations were carried out to the overhead girders, or arches, of the aqueduct and caissons so to allow the passage of larger boats, with a capacity of 40 tons instead of the previous 25 tons limit. As recorded in the minutes of the Trustees:

'The North Staffordshire Railway and Canal Company have intimated that they were widening their canal, (between Middlewich and Anderton), by which they anticipated an increased traffic would be forwarded via the Anderton Lift, and requested us to raise the Cross Stiffeners above the Caissons so as to admit the passage of larger boats. We agreed to their request, and the alterations, which were estimated to cost about £150, will be completed next month.'

By 1893, Saner was sufficiently concerned about the strength of the corroding rams that he began investigating alternative methods of extending their working life. He was especially worried about the condition of the elements of the rams which were below the bed of the lift basin and hence could not be inspected. One of his ideas involved the use of counterweights alongside the caissons in order to reduce the working pressure on the rams. For whatever reason, however, this was not progressed.

The circular process of grooving of the rams, filling with copper strips, followed by bi-metallic corrosion and hence more grooving, was accelerating year by year. Following experiments carried out at the nearby laboratories of Messrs. Brunner, Mond & Co., Saner presented a special report to the Trustees in January 1895, which set out the extent of the problem, as follows:

'The grooving action at Anderton commenced, as far as I am able to ascertain, about the year 1882, when the first pieces of copper were put in – this was seven years after the lift first commenced to work; and in 1889, soon after I took charge, I found that there were already some 246 lbs. of copper strips, averaging about 3/8 inch broad by 1/4 inch thick. The action at that time was not very rapid, but fresh strips had to be put in from time to time, and I found, as will be seen from my reports, that the necessity arose more frequently, until now there are nearly as ascertainable 783 lbs. of copper in the two rams. On careful examination it appeared to me that the decomposition took place most rapidly in proximity to the copper already in, and this view was strengthened by some of the pieces becoming loose and having to be replaced with broader strips . . . . The iron affected and in close proximity to the copper was soft and friable, and had every appearance of the action being electrolytic in its nature.'

Saner concluded that:

'. . . . distilled water, that is water which has been converted into steam and then cooled again into water by means of a condenser, being apparently the only solution of the difficulty.'

He then suggested two ways of proceeding:-

'The first is to leave everything as it is and fill up the fresh grooves as they form with copper, until the rams become at no distant date so weakened that they would have to be withdrawn and replaced with new ones. This would mean a long stoppage of the traffic, and a probable expenditure of £800 to £900 each ram, also until the new rams were in, there would be an increasing difficulty in keeping the 'packing' from being torn by the grooves, and the risk of inability to retain the pressure water in the cylinders . . . Or, secondly, to put down plant for purifying the River Water,

and prolonging the life of the rams for many years.

I propose the latter method and would obtain it in the following way:

(a) Work the boilers to their full pressure of 70 lbs. per inch.

(b) Convert the engines into compound, by removing one cylinder and replacing it with a larger one to act as a low pressure cylinder.

(c) Construct a surface condenser to not only act for the exhaust steam from the engines, but also to be arranged for working independently and condensing steam direct from the boilers.

(d) Storing the water so condensed in large tanks, placed at a suitable level so as to supply the pumps with less 'shock'. These tanks would be made from old boilers I have in stock.'

Saner suggested that the cost of the second option would not exceed £590 and would have the dual benefit of prolonging the life of the rams and reduce operational costs by economising on the use of coal. He concluded his report with the following warning to the Trustees:

'Whether the quality of the canal water has deteriorated more rapidly since the erection of the chemical works at Lostock, I am unable to say at present for want of comparative analysis, but whatever the cause the sooner the use of that water for the lift purposes is discontinued the better, and I am convinced that the suggestion I have put before you is the most practical way of dealing with the matter.'

All in all, the report written by Saner was lucid and well presented. It would seem that the Trustees had little option but to accept their Engineer's advice, particularly as the income of the Weaver Navigation Trust had come to depend so much on the traffic passing through the lift. Notwithstanding this, however, the Trustees chose to defer any decision to a later date due to the large cost of the works and the financial position of the Trust at the time.

In his end of year report to the Trustees just five months later, Saner recorded that:

'By re-arrangement of the working staff at the Anderton Lift, a saving of wages has been affected.'

and then rather pointedly:

'The grooving of the rams continues with increased severity, and there is no doubt something must shortly be done to prevent its further extension.'

Two years later in 1897, the Trustees took the inevitable decision, and the works to convert the hydraulic system to distilled water were carried out. In his 1910 report, Saner stated that, in his opinion, the conversion to distilled water prolonged the life of the rams by as much as ten years:

'There is no doubt that, had the electrolytic action been allowed to continue at the same speed, reconstruction of the lift, (which eventually took place in 1908), would have had to be undertaken much sooner.'

★ ★ ★ ★ ★ ★ ★ ★ ★ ★

In 1896, a new problem arose relating to the stability of the bank at the lift site which supported the Trent & Mersey Canal. As recorded in the minute book of the Trustees:

'They, [the Anderton Committee of the Trust]*, afterwards visited Anderton, and inspected the wall that had moved on the East side of the Basin and which is in close proximity to the Hydraulic Lift, and they are of the opinion that a wall as suggested by the Engineer should be built outside of it immediately to ensure the safety of the Lift, otherwise a disaster of great magnitude must result, as the side of the Trent & Mersey Canal may give way at any moment, and entirely stop not only the navigation of the Canal, but also that of the River Weaver. The building of the said wall would necessitate the expenditure of £1,000 to £1,500 .... They attribute the cause of the wall moving to the percolation of the Brine from salt stored from time to time in Mr. Thompson's Warehouse as well as to the extra weight placed on the side of the hill, and it is absolutely necessary that this Warehouse should be removed forthwith under clause 7 of the Agreement with Mr. Thompson .... They also inspected the Easterly salt shoot, and are of the opinion that it should be almost entirely renewed, and the position slightly altered in consequence of the proposed new wall.'*

The full extent of the repairs carried out are revealed in Saner's Annual Report to the Trust for the year ending 28 November 1897:

'As the movement was close to the Foundations of the Aqueduct to the Lift, any further extension would certainly have endangered that structure. I considered it necessary therefore, to have the matter attended to at once, and in view of the important issues at stake, I asked the Trustees to call in Sir E. Leader Williams (a previous Engineer to the Trust) for consultation, and after he inspected the locus in quo it was decided to partially fill in the Basin and drive piling on lines shewn on plans submitted to the Trustees .... While the work was in hand it was found that a re-arrangement of the Salt Shoots and the Landing Stages would be necessary, and this was carried out in accordance with the plans submitted to the Trustees.'

And so it was that the looped basin, which had been developed some fifty years earlier to enable better access for boats from the River Weaver to the salt chutes, the landing stages and the warehouses, was reduced to two separate arms, one to the east and the other to the west of the lift. The overall cost of the changes to the Weaver Trust was £2,436-7s, of which £1,250 was claimed back from Mr. Thompson, the owner of the warehouse that was purported to have caused the collapse of the basin wall.

★ ★ ★ ★ ★ ★ ★ ★ ★ ★

Problems were still being encountered with corrosion on the lift. In 1898, weakening of one of the flanges of the main equilibrium valve between the two rams was identified. Saner, was at that time '*on the Continent*' and so it was left to his assistants to make the decision to take one of the caissons out of operation for a number of days, to permit repairs to be undertaken. This again resulted in

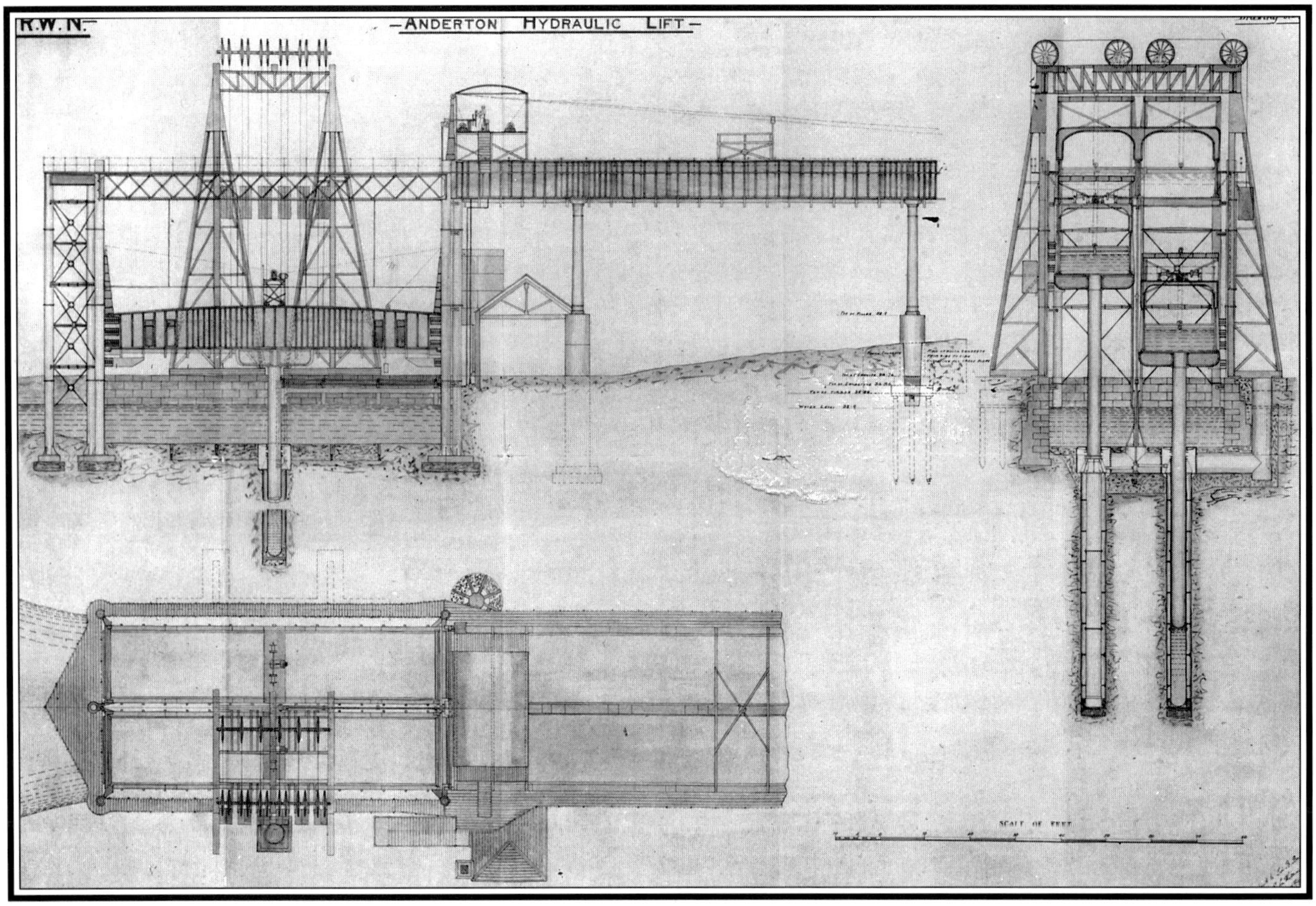

Drawing of proposed modifications to the hydraulic lift, circa 1893, which were never carried out. It would appear they were intended to reduce the load on the weakened rams by means of counterweights and wire ropes. Courtesy The Boat Museum, Ellesmere Port

single-caisson working of the lift until the repairs had been completed. Interestingly, in his Annual Report in August 1898, Saner explained that:

'Owing to the incident occurring two days before what is known as Stoke 'Wakes', the traffic to be passed with one 'caisson' was exceptionally heavy, and involved a large amount of overtime. I have to highly commend the staff in charge of the Lift, and the Mechanic's staff, for the successful and energetic manner with which they dealt with it.'

During the closure, the opportunity was also taken to procure a new valve, together with replacement pipework.

In the same report, Saner attempted to justify the increasing expenditure in repairing the lift by reminding the Trustees of the importance of the structure to the well-being of the Weaver Navigation as a whole:

'As the expenditure on this Lift has been remarked upon, I should like it to be remembered that though the structure is now about 23 years old, and naturally requires somewhat more attention than formerly, on completion of contemplated repairs the expenditure will again be reduced. It should be borne in mind that not only the direct, but also the indirect, benefits to the Navigation in the shape of the Pottery traffic are only made possible by its existence, and that without the extension of the River to the Midlands, the Anderton route is the only route which can be relied on for the development of traffic outside the Northwich and Winsford traffic.'

What Saner did not say, presumably because the figures were already available to the Trustees, was that total tonnages of cargoes being passed through the lift were higher than ever, having taken a marked upturn after 1895. The cargo carried in 1898 was just over 186,000 tons, which in crude terms represented about 7,500 laden vessels or on average about 25 laden vessels a day. It is facts like this which demonstrate that the Anderton Basin was a very busy commercial junction in the late 19th century. The resulting income to the Trustees, both direct tolls for the use of the lift and indirect tolls for the additional traffic that it attracted to the Weaver Navigation, was not insignificant. The increasing trade associated with the lift was, however, occurring against a gradual decline in income for the Trust from the Weaver Navigation as a whole. In 1881, the annual income from the Weaver Navigation had peaked at £70,105, but the following years generally saw reductions, until by 1895 the income was less than £50,000. It should be remembered that the reduction in income was not entirely a result of a loss of traffic but also a reflection of the reduced tolls being charged in order to survive against increasing competition from other forms of transport,

THE DEVELOPMENT OF ANDERTON BASIN 1876 - 1900

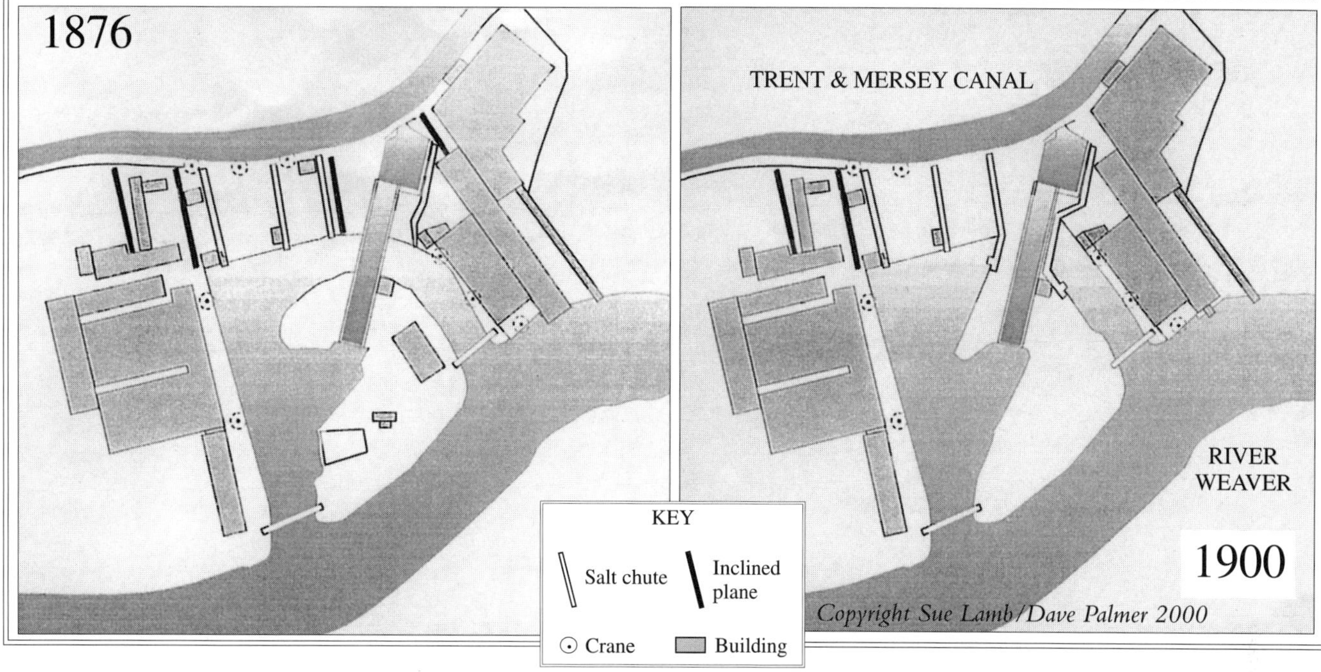

The narrowboat Euston, *belonging to John Walker of Stoke-on-Trent, is seen here about to ascend using the western caisson sometime in the 1920s. One of the lift operators keeps an eye on progress. This close-up photograph, taken from almost underneath the footbridge across the channel into the lift, shows the caisson gate and its operating pulleys in good detail.* Courtesy The Boat Museum, Ellesmere Port

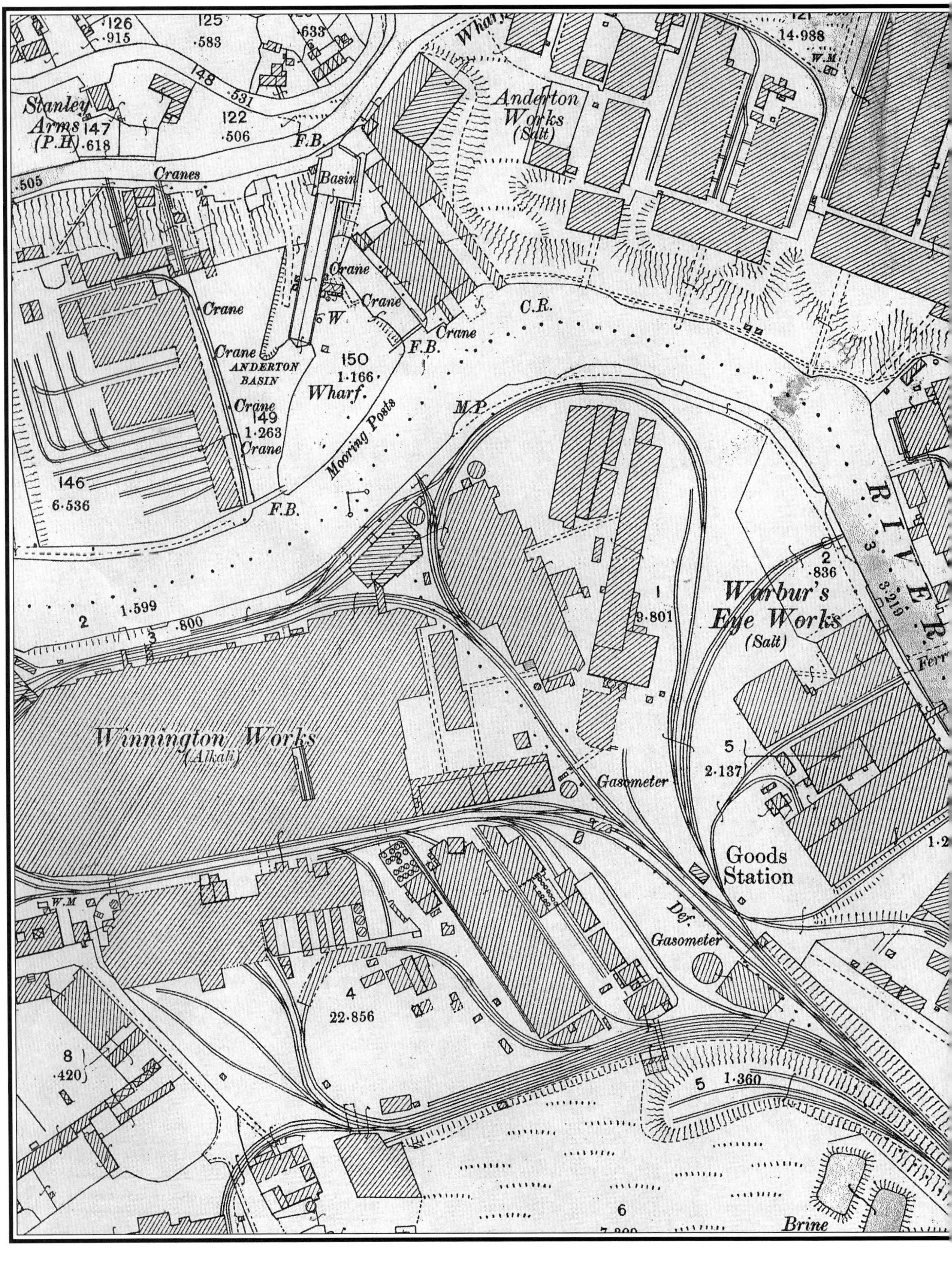
Stanley Arms (P.H.)
Cranes
Basin
Crane
Anderton Works (Salt)
C.R.
F.B.
ANDERTON BASIN
Wharf.
Mooring Posts
M.P.
Warbur's Eye Works (Salt)
RIVER
Winnington Works (Alkali)
Gasometer
Goods Station
Def.
Brine
W.M
146
6.536
149
1.263
150
1.166
2
1.599
.800
4
22.856
8
.420
5
1.360
6
1
9.801
5
2.137
2
.836
3
3.219
122
.506
147
.618
126
.915
125
.583
148
.531
14.988

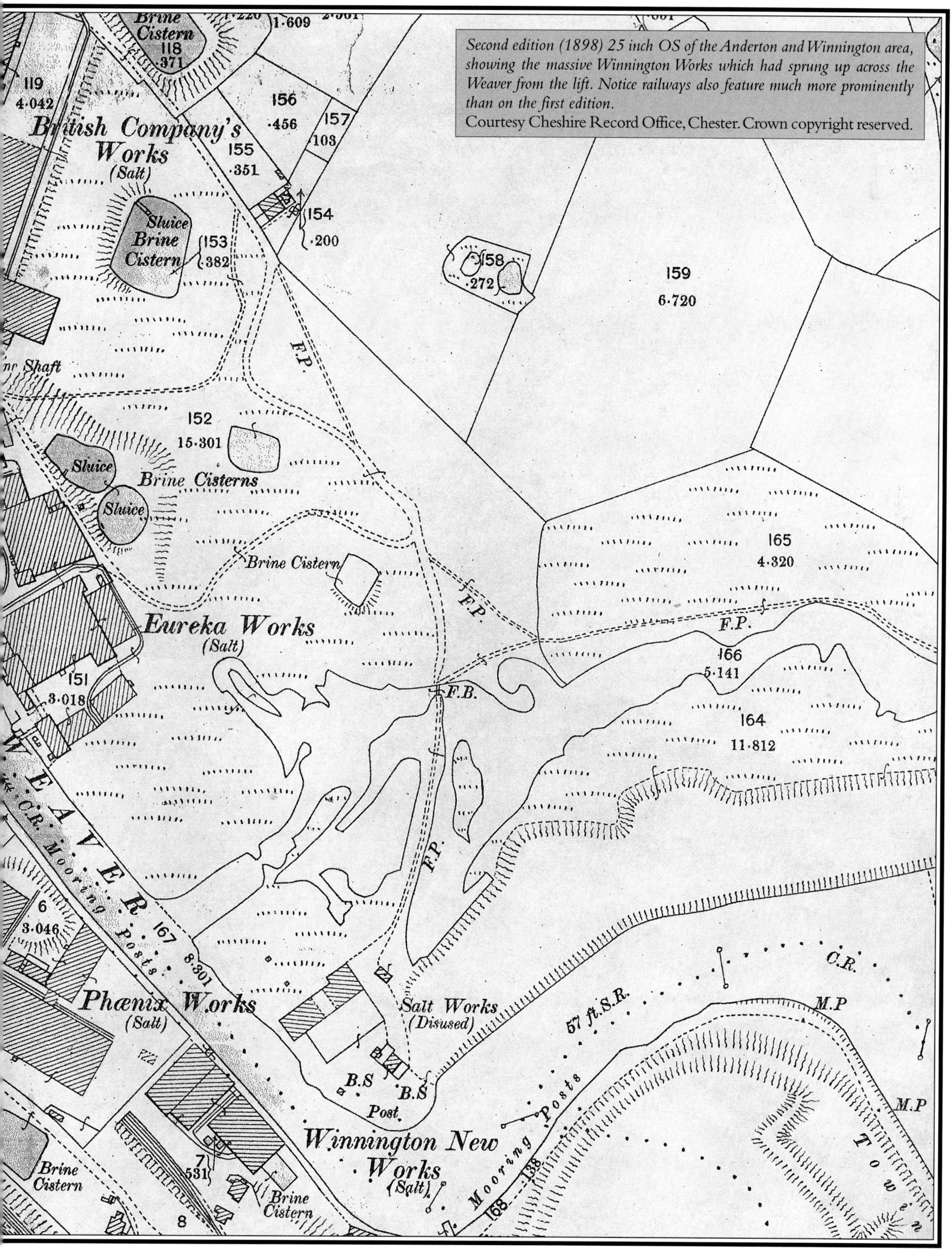

Second edition (1898) 25 inch OS of the Anderton and Winnington area, showing the massive Winnington Works which had sprung up across the Weaver from the lift. Notice railways also feature much more prominently than on the first edition.
Courtesy Cheshire Record Office, Chester. Crown copyright reserved.

This photograph of the hydraulic lift was taken circa 1906, shortly before the commencement of works to convert the lift to electrical operation. Note the larger lift control cabin visible on the platform above the aqueduct and also the rail lines either side of the hut; these do not appear in any of the previous views and almost certainly had been recently installed for a steam crane which was used during the conversion work (see page 96). On the far left, salt cascades out of a chute into the hold of a waiting vessel.

Courtesy The Boat Museum, Ellesmere Port

primarily the railways and pipelines.

★★★★★★★★★★

Despite the conversion to distilled water, difficulties continued to be encountered with respect to corrosion. On Wednesday 14 June 1899, yet another incident occurred. Saner advised the Trustees of the problem:

'. . . about 5 p.m. some of the connections conveying the pressure water, gave way. It was found possible, without serious risk, to work the traffic until Saturday, when I gave orders for the repairs to be taken in hand. It was found not possible to complete by Monday, so the traffic was stopped until Thursday morning. Some very pressing Corn cargoes and Bag Salt cargoes were transhipped by means of derricks, and everything possible done to minimise inconvenience to the traders.'

In August of the same year, traffic on the lift was stopped for a period of a little over four days to allow essential maintenance works to be carried out. Saner reported:

'In accordance with my notice, this Lift was stopped from 4.0 p.m. on the 6th instant to 6.0 a.m. on the 11th instant, during which time the Upper Basin and Aqueducts were thoroughly cleansed, the walls pointed and grouted with cement, and the ironwork scraped and compositioned. The accumulator was emptied of ballast and thoroughly overhauled. A large number of the hydraulic and steam pipes and valves were also either renewed or repaired.'

Rather surprisingly, given the poor condition of the hydraulic rams, Saner was able to conclude that:

'I am pleased to report that the main structure, so far as examined, was found in very good order.'

The following year, the lift was once again closed to traffic for maintenance and repair, this time for nominally 6 days. During this period the lower basin was drained and cleared of sediment; the supports for the caissons were adjusted; the engine cylinders and lifting chains overhauled; the wrought iron gates to the caissons were repaired; and the whole of the ironwork to the caissons, both inside and out, scraped and painted with tar. In addition, the troughs for carrying away the surplus water from the lift were renewed and joints, not normally accessible, refurbished. Saner concluded his report to the Trustees by saying:

'I have to report that the machine as a whole is in good condition, but there are numerous minor details requiring attention as it becomes possible.'

It is almost certain that, in order to minimise the stoppage period, the maintenance and repair works carried out involved several teams of men, more than likely working round the clock.

In January 1901, electric arc lamps were installed for the first time, presumably to enable the lift to be operated for longer periods during the short winter days. In August further repair work was done, which included overhauling the engines to the accumulator and the installation of timber boarding at the northern end of the lift to prevent water and ice from falling onto the engine house. Again, the ongoing task of filling grooves with copper, this time on the up-river ram, was carried out. Saner recorded that by this time:

'the grooving had extended downwards, the lift had to be stopped in order to take off the face joint and get the ram higher up than usual, to obtain access to the affected part.'

In his Annual Report in November 1901, Saner wrote:

'the Anderton Lift has had a thorough overhaul and repair, (during the year), and is now doing more work than ever before.'

In September 1902, Saner drew up a specification for the adoption of electricity at the lift. This related to the use of electrical power to replace some of the manual tasks on the hydraulic lift and was not a precursor to the total electrification that was to follow a few years later. These works included the conversion of the vertical lift gates on both the aqueduct and lift structures to electrical operation, thereby enabling the workforce on the lift to be reduced by three, together with the replacement of the steam engines for the accumulator by electric pumps. The valve house at the top of the lift was also converted to electricity. The steam engines were, by that time, in poor condition having been in almost continuous operation since 1875 and Saner had assessed that it would be more cost-effective to convert the accumulator system to electricity than it would be to replace the steam engines. He recorded in his 1910 report that:

'. . . this proved to be so; in spite of an increase of nearly 50 per cent in traffic, the cost of electric power was under 10s. a day, while the cost of coal had formerly been 12s. to 14s. according to the market price.'

Around the same time, during routine maintenance of the structural elements of the boat lift, severe corrosion of the wrought iron cross beams beneath the floor of the aqueduct was detected. Repairs were subsequently put in hand with additional cross beams being installed to further strengthen the aqueduct.

Corrosion of the rams and pipework continued to be a problem. The poor condition of the hydraulic system, with the need for frequent repairs often with prolonged stoppages to traffic on the lift, was becoming a major concern to Saner. He recorded that, by 1904, the structure was making about 140 lifts a week and passing over 15,000 boats a year, with some 190,000 tons of cargo. Saner undoubtedly drew his concern to the attention of the Trustees on a number of occasions and at one stage reported:

'. . . as the parts of this Lift, more especially the main rams and cylinders, are all suffering from the action of water (vide my Reports from time to time) I think that serious consideration should be

This photograph appears around the same time as the view on page 88 and, if so, was taken shortly before work began on reconstruction of the lift in 1906; the new design required more space and the salt chute on this side had to be removed. It may be that dismantling of the chute is just beginning, the timbers retrieved being loaded into the barge. Alternatively, it is an earlier date and the salt chute is being repaired. Note also the caissons in mid lift.

Courtesy The Boat Museum, Ellesmere Port

given to this question, as any stoppage of the Lift would mean a serious detention to and loss of traffic.'

And again on 25 March 1904:

'It will be remembered that these (the rams) were bad some years ago, but that they have been preserved by the use of special water, and although I hope it will be some little time before it is necessary to undertake the renewal of the rams, yet it is a matter which should not be lost sight of.'

It was not lost sight of, for, just a few months later, Saner prepared a comprehensive report to the Trustees, dated 17 August 1904. By this stage, he had come to the conclusion that something more radical was required if the lift was to be kept in service. In the report, Saner firstly reminded the Trustees of the original reasons for building the lift and its increasing commercial importance to the Weaver Navigation as a whole. He stated:

'. . . the traffic, (from the Potteries to Weston Point via Anderton), has, during the last decade or so very much increased, and during the recent decline of the Salt Trade, formed a most valuable and welcome addition to the revenues of the Trust. The tonnage last year passing over the Lift amounted to 188,000 tons, against 60,000 in 1890, and the present revenue from such tonnage is therefore very considerable. There were 6,981 operations of lifting last year.'

He then described the full extent of the repair works that would soon be necessary to the lift. These included renewal and strengthening of the bottom plating of the aqueduct and wrought iron supporting beams; replacement of two of the aqueduct vertical-lift gates and repair of the other two; refurbishment of timber stairways and galleries; masonry repairs; renewal of wrought iron stiffening girders which had been exposed to splashing; repair to the bottom plating of the caissons; renewal of three of the caisson vertical-lift gates; renewal of the cylinder to the hydraulic accumulator and the provision of better guides; and finally replacement of the pressure pipes between the rams. He then reminded the Trustees of the very poor condition of the hydraulic rams and expressed uncertainty regarding the integrity of the hydraulic cylinders sunk into the ground beneath the caissons. In Saner's own words:

'It will be seen, therefore, that considerable repairs must presently be done if this excellent machine is to continue to meet the requirements of traffic.'

He continued by suggesting two ways of proceeding, whilst taking account of the need to minimise disruption to the flow of traffic through the lift. The first option was to build a second lift between the two waterways, either 250 yards downstream of the original lift or at Whatcroft, several miles to the south of Northwich, the latter requiring either a new cut from the Weaver or the use of the River Dane. The second option was to '*adopt some other form of mechanism which would enable the Lift to be rejuvenated*'. Saner continued by describing the details of each of the options and highlighting their advantages and disadvantages. With respect to the option to rejuvenate the lift, he summarily discounted the possibility of carrying out the repairs whilst retaining hydraulic operation, as it would cause unacceptable disruption to traffic. He then unveiled his radical proposals for the conversion of the lift to electrical operation. Again, in his words:

'. . . but I have found a solution which would be satisfactory, but which at the same time does away with the hydraulic power and submerged works entirely.'

And so, the Special Report to the Trust in August 1904 signalled the beginning of the end of the hydraulic boat lift.

INFORMATION SOURCES FOR CHAPTER 6

Reconstruction of the Canal-Boat Lift on the River Weaver at Anderton. *Paper by J. A. Saner in* The Proceedings of the Institution of Civil Engineers, Volume 180, Part 2, 1910.

River Weaver Navigation. *Report of Mr. J. A. Saner, Assoc. Mem. Inst. C.E., on the corrosive action which is taking place in the main hydraulic pressure rams of the Anderton Lift, 30 January 1895.*

River Weaver Annual Accounts and Reports – 1881 to 1895.

Continental River and Canal Works. *Paper in The Proceedings of the Institution of Civil Engineers, Volume 96, 1889.*

Minutes of the Proceedings of the Weaver Navigation Trustees 1896 to 1904. *British Waterways' Regional Office, Northwich.*

A view at ground level across the lift basin with the erection of the new steel 'A' frames ongoing in the background. The lift is still being operated by means of the hydraulic rams, one of which can be seen fully extended in the left foreground, whilst the eastern caisson is at its lowest position within the wet well. It is interesting to note the gantry in front of the caisson supporting the pipework running between the rams and the control cabin. Also of interest is the gantry in the middle of the eastern caisson, on which is mounted the motors for the caisson gates, installed in 1902. Courtesy The Boat Museum, Ellesmere Port

CHAPTER 7
CONVERSION TO ELECTRICAL OPERATION

'My proposal, therefore, is to do away altogether with the hydraulic power, erect a network of steel girders over the entire present structure, and by means of wire ropes, supporting at one end the caisson, and at the other a balance weight, and passing over a multiplicity of pulleys, similar to the arrangement employed by the Manchester Ship Canal Company at the Weaver Sluices, lift or lower the caissons by means of an electro-motor geared to the aforesaid pulleys.'

Extract from a *Special Report to the Trustees* by J. A. Saner, 17 August 1904

The above proposal for the conversion of the Anderton Boat Lift from hydraulic to electrical operation was made to the Trustees of the Weaver Navigation by their Engineer, J. A. Saner, almost thirty years after the commissioning of the original lift in 1875.

In most respects the original structure had been a great success and had commercially more than justified its cost of construction. The much needed improvements to transport links, achieved for local industries to the outside world by the presence of the lift, were the cause of a significant upturn in the fortunes of the Weaver Navigation. Against this commercial success, however, the original boat lift was plagued throughout much of its thirty year life by corrosion, especially to the hydraulic rams. The condition of these rapidly worsened over the latter years and undoubtedly shortened the working life of the hydraulic lift.

The Trustees faced a real dilemma; how to carry out the essential repairs or replacement of the lift, whilst maintaining the valuable income brought about by its working. The Trustees, no doubt, were all too aware of the major downturn in income that resulted from the six month closure of the lift when a hydraulic ram failed in operation in 1882. With the annual tonnages in the early 1900s being of the order of five times higher than those in the early 1880s, the impact of any prolonged closure to the lift for refurbishment would have been devastating to the Trustees. It was at this point that Saner came to their rescue by recommending conversion to electrical operation. In his special report, which is quoted at the beginning of this chapter, he extolled the benefits of his proposals for converting the lift to electrical operation:

'This design has many advantages over the hydraulic arrangement at Anderton:

(1) Either caisson can be worked independently of the other.

(2) The depth of water is uniform, and no water is abstracted on the upward movement.

(3) All the complicated hydraulic pipes, with the corrosion, liability to freeze and burst, accumulator, engine room staff, and boiler repairs, are done away with.

(4) The number of attendants can be reduced one man even below the present reduction.

(5) All the important parts are above ground, and can be examined without stoppages or inconvenience.

(6) There is absolutely nothing to get out of order except the driving motor, and by having a spare armature the delay from this would be exceedingly small and of rare occurrence.

(7) The cost of construction will turn out, I have every confidence in stating, less than that of any other system.

I had originally no intention of suggesting this arrangement for Anderton, but after considering the delay necessary to repair the present rams, etc., and the fact that the new system could be applied with only sufficient stoppage to actually connect the ropes, alter the gates and a few other details, probably 3 stoppages of say 2 weeks each, I felt justified in putting it forward.'

Under the circumstances, Saner's assertion that the conversion works could be carried out with just three relatively short stoppages must have been music to the ears of the Trustees. This and the other assertions by Saner may have appeared to be extravagant but, with the exception of item 6 perhaps, he was shown by subsequent events to be correct. The next statement in the special report further demonstrates his total confidence:

'I am quite willing, in view of the importance of the matter, to confer with any competent Engineer before commencing the work, if it is thought expedient.'

The report concluded with estimates of costs for the various alternatives. His estimate for the electrical conversion was as below:

SCHEME No. 4
CONVERT ANDERTON TO ELECTRICALLY-DRIVEN EQUILIBRIUM LIFT

Equilibrium Lift. (Only 6 weeks stoppage, in probably 3 separate fortnights)

New Work	£12,894
Repairs	£ 2,100
	£14,994

This estimate compares very favourably with the other options being considered at the time. Saner advised that the likely cost of repairing the hydraulic lift, with a long stoppage, would be between £21,600 and £23,100, depending upon the eventual cost of the new rams and allowing for transferring cargoes by other, more costly, means during the stoppage. His lowest estimate for a complete replacement lift, which would have been located at Whatcroft, was £105,334.

From the evidence presented by Saner, the Trustees had little choice but to accept his proposals; to proceed with the electrical conversion was clearly the low risk option with the lowest capital expenditure. Furthermore, it may have been on the Trustees' minds that, should it later prove necessary or appropriate, a replacement lift at an alternative location could still be built.

The final suggestion made by Saner to the Trustees was as follows:

'The design is one which I have for other purposes, and without thought of Anderton, been working at for nearly 12 months, and have at my own expense made a model, and taken out Patents; the model I have here on view, and the Patent rights I shall be prepared to make arrangements about, if the design is adopted.'

Whether this move was as audacious in 1904 as it sounds today, I cannot say. It is simply recorded in the minutes of the Weaver Trustees that:

'The Anderton Lift Committee discussed the question raised by Mr. Saner re his remuneration in regard to his patented method of converting the hoistage from hydraulic to electrical motor. Mr. Saner was awarded a sum of £1,050 to be spread over 3 years.'

In this day and age, however, it does seem rather tenuous for Saner to assert that his ideas had been developed without consideration to the ailing boat lift at Anderton and, by implication, separate from his professional obligations to the Weaver Trustees. Perhaps, however, his actions were viewed differently at the beginning of the century.

Notwithstanding the above, the Trustees took up Saner's suggestion of having the design verified by a third party and subsequently appointed Sir Benjamin Baker to carry out a detailed review of the proposals. Subject to some modifications, Baker approved Saner's proposals and as a consequence the Trustees set in motion the process of raising the necessary monies. In the meantime, Saner was instructed to proceed with the preparation of detailed drawings and other necessary documentation.

One of the main modifications made as a result of Baker's review was the adoption of safety ropes to address the potentially hazardous consequences of a slippage of the wire ropes over the pulleys. Saner had initially proposed the simple arrangement of passing all the wire ropes directly between the caissons and the counterweights over the pulleys. Baker, however, advised that a number of the ropes should provide a 'positive drive' by being connected directly to the pulleys, thereby reducing the likelihood of slippage in the event of a weight differential developing, for whatever reason, between the counterweights and the caissons .

Subsequently, in a further report to the Trustees on 10 November 1904, Saner advised that:

'With regard to the future working of the Lift, the new arrangement will have so few working parts that I anticipate the cost of the repairs will be very considerably reduced, especially when the aqueduct and caisson gates have been thoroughly overhauled, a large portion of the last two years expenditure being in gradually renewing and repairing these gates . . .

Under the new arrangement, the capacity of the Lift will be increased, as each trough will be independent, and the caissons can be worked according to the exigencies of the traffic. For instance:- At certain parts of the day most of the traffic is going down-along, and at other parts of the day it is passing up-along, more especially is this the case on Mondays and Saturdays. Under these circumstances, both caissons can be raised to the top, or lowered, without waiting for alternative lifts.

There will also be a saving in time, through the water level in the caissons being constant, instead of having to be varied, as now, but this will be somewhat counteracted by there having to be additional gates to lift at the lower end when the caissons are down.'

With hindsight, it could be said that, in many respects, Saner was overstating (whether deliberately or not) the perceived benefits of the proposed conversion to electrical operation. For example, it is difficult to understand how he could claim that the proposed lift had fewer moving parts than the original lift. The hydraulic lift contained very few moving parts, which included the rams, the control valves, the steam accumulator and the guillotine gates to the aqueduct and caissons, (i.e. 2 aqueduct gates and 4 caisson gates). In contrast, the proposed conversion would have the operating headgear, to include electric motors, gearboxes, drive shafts, pulleys, wire ropes etc., as well as two additional gates at the river entrance to the lift.

Saner may, to some extent, have been justified in claiming speedier operation for the converted lift. The time taken to move the caissons from one level to another by hydraulic operation was 8 minutes, whilst under electrical operation it would be two minutes quicker. However, as pointed out by Saner, any time saved by avoiding the need to make the final adjustment to the levels of the caissons on the original lift would almost certainly be offset by the need to operate two sets of gates at the river entrance of the converted lift.

It was recognised at the time that the construction of a new lift at a new location would have been a relatively simple, albeit costly matter, and that the cheaper option of converting the existing lift to electrical operation posed several significant problems. These were detailed in a report entitled *Anderton Boat Lift – 1908*, which was published in *The Engineer* on 24 July 1908, and presumably written by Saner on behalf of the Weaver Trustees:

'(a) Non-interference with the traffic;
(b) Future accessibility for repairs;
(c) Cost of carrying out the works sufficient to make the machine thoroughly safe again.'

There is little doubt, however, that the most difficult task facing Saner, with respect to the proposed conversion, was that of minimising disruption to the valuable traffic using the lift. By all accounts, on this matter at least, his efforts were to be very successful.

★ ★ ★ ★ ★ ★ ★ ★ ★ ★

During 1905 the North Staffordshire Railway Company, which by that time had taken over the running of the Trent & Mersey Canal, wrote to the Trustees asking for consideration to be given to the installation of different size caissons during the conversion works. Their proposal was for one of the caissons to be widened by as much as 5 feet 6 inches to 20 feet and the other reduced in width accordingly. By such means, the lift would be better suited to the traffic then on the canal, which included a significant number of wide barges as opposed to narrow boats. This suggestion, however, was not progressed, presumably because of the additional cost. There is no record of any response given to the Railway Company on this matter.

★ ★ ★ ★ ★ ★ ★ ★ ★ ★

It was not until early 1906, however, that the Trustees gave the go ahead for orders to be placed for the conversion of the lift to electrical operation and preparatory work began almost immediately. It is interesting to note that, unlike the works to construct the hydraulic lift, which was achieved by means of a single contract with Emmerson Murgatroyd & Company, the Trustees elected to have the conversion works carried out principally by their own workforce under the management and supervision of Saner, their Engineer. This decision by the Trustees displayed enormous confidence in his abilities and those of the Weaver staff.

One of the first tasks to be carried out was the alteration of the lift well and strengthening of the foundations. This work was subsequently described in the *Anderton Boat Lift – 1908* report, as follows:

'An essential feature of the new design is the provision of a dry dock for the caissons to descend into, (the lift well for the hydraulic lift being filled with river water), and as the existing dock had no bottom, but merely (masonry) walls at the sides and one end, it was necessary to put in an inverted arch and end river gates, which not only made a dry basin, but also formed a support for

The next few pages contain a series of photographs showing the conversion of the lift to electrical operation. Due to the complexity of the conversion works, the photographs do not necessarily follow in strict chronological order but have been arranged to show the general sequence of the conversion works. The view above shows excavation work in preparation for the later construction of the concrete counterfort walls, as foundations for the new 'A' frames along the eastern side of the lift. One of the caisson river end gate frames is partially visible on the extreme right of the photograph.
Courtesy The Boat Museum, Ellesmere Port

This photograph shows the works to convert the wet well to a dry well, which was undertaken during a 19 day stoppage to traffic on the lift during April 1906. During this period, work was progressed on a 24 hour basis by relays of men. In this view, excavation work can be seen underway, whilst a recently concreted section of the floor appears in the foreground. Courtesy The Boat Museum, Ellesmere Port

A second stoppage, this time of 10 days, was required during August 1906 to enable the completion of the work affected by water. This included the installation of the new river basin gates and frames, as shown here. A sailing packet can be seen in the river basin in the left background. There were no further stoppages until April 1908, close to the completion of the conversion work. Courtesy The Boat Museum, Ellesmere Port

Right: *Another view of the counterfort wall work for the 'A' frame foundations along the eastern side of the lift. The below ground concrete work appears to be complete and the cast iron baseplates for the above ground steelwork are being installed. To the right of this view, note the steam operated crane on the rails previously seen in the photograph on page 88.*

Left: *Another photograph of the steam crane, here lifting one of the massive granite blocks that were to be built into the top of the side walls of the dry well and the counterfort walls. The cast iron baseplates for the vertical columns of the 'A' frames were subsequently anchored down to these blocks.*
Left below: *One of the granite blocks being lowered into its final position, seemingly onto one of the concrete counterfort walls.*
Below: *Workmen pause momentarily for the camera, during what is probably the excavation of the chamber, on the eastern side of the lift, which would eventually house the electrical pump to keep the dry well drained.*
All courtesy The Boat Museum, Ellesmere Port

Above left: *One of the new steel vertical columns being lifted into place on the eastern side of the lift, alongside one of the 1875 cast iron guide columns at the aqueduct end of the lift. The below ground concrete works have been completed and the new column will be bolted to a cast iron baseplate like the one in the foreground.* ***Above right:*** *The same view a few days later, with the vertical column now securely in place and the lower sections of the angled bracing column of the 'A' frame being installed.* Both courtesy The Boat Museum, Ellesmere Port

A group of Weaver Navigation employees pose cheerfully for the camera during a break in work on the conversion of the boat lift. Whilst the materials and equipment were ordered in, the conversion work on site was carried out by the Weaver Navigation Trust's own men, under the direct supervision of Col. J.A. Saner.
Courtesy The Boat Museum, Ellesmere Port

Rivetters at work inside one of the caissons, carrying out extensive repairs to the floor plates which were necessary due to their prolonged exposure to the badly polluted river and canal water. The lad on the left has just extracted a red hot rivet from the brazier. A tarpaulin has been slung across the caisson, presumably to protect the brazier and those working inside from water splashes.
Courtesy The Boat Museum, Ellesmere Port

Each caisson in the 1875 hydraulic lift was raised and lowered by a single, centrally positioned ram, with the load being transferred to the underside of the caisson by means of this cruciform support structure. During the third and final stoppage of 20 days in April and May 1908, the cruciform supports were removed and the caissons modified, prior to their being attached to the wire ropes and counterweights in readiness for electrical operation. This photograph shows preparatory work taking place, before the removal of one of the cruciform support structures. Courtesy The Boat Museum, Ellesmere Port

Another view of the cruciform support structure, after the removal of the caisson. In the background, the new 'A' frames and diagonal ties are in place, with the cast iron counterweights and wire lifting ropes lined up behind in readiness for connection to the modified eastern caisson.
Courtesy The Boat Museum, Ellesmere Port

Left: *The steel guides for the new river gates in the dry well. The original cast iron guide column is visible in the left foreground. A temporary steel gate is holding back the water from the river basin, whilst workmen, including a diver in full gear, pose for the camera in the background.*

Opposite page, top: *A view of the front of the lift, which gives a good idea of the scale of the task facing Saner and his workforce in converting the lift to electrical operation, whilst minimising disruption to traffic through it. The nearest of the new 'A' frames is in place on the eastern side of the lift, while others are being installed behind. Note the derricks mounted on top of the wooden scaffolding on either side of the lift.*

Left: *A view of the conversion works from the west, across the river basin, with temporary wooden scaffolding in front of the recently installed 'A' frames. Note that the wooden salt platform visible in the photograph on page 90 has been removed to make way for the scaffolding. The focus of the work has now moved to the upper level, in readiness for the installation of the five steel cross beams for the machinery deck.*

Opposite page, bottom: *During one of the stoppages, the opportunity was taken to drain the canal basin to allow it to be de-silted, with the bed being lowered and repuddled. The interior. of the aqueduct is visible behind and Winnington Works peers through the background haze.*

Left: *A steam derrick was erected on top of the wooden scaffolding to assist with the installation of the steel beams of the machinery platform. The new 'A' frames are visible on both sides of the old lift structure. Work has briefly come to a halt as the men pose for the photographer, who would appear to have positioned his camera on one of the adjacent salt chute platforms.*

All courtesy The Boat Museum, Ellesmere Port

the side walls, upon which the upper structure, weighing some 1676 tons dead load, was to rest. The foundation was only fair, being gravel mixed with a small amount of sand, so that it was necessary to spread the weight over a sufficient area and make the mass perfectly solid and water-tight . . .'

This foundation work was carried out on a 24-hour basis, with relays of men, during a 19 day stoppage of the lift between 12 and 30 April 1906.

A second stoppage, this time of 10 days from 4-13 August 1906, saw the completion of the preparatory work that was affected by water, which included the installation of the wrought iron frames for the new river gates.

There were no further stoppages to traffic between September 1906 and April 1908. During this period, the conversion work was progressed with the construction, below ground, of the concrete counterfort walls for the 'A' frames along each side of the lift basin. This work could be carried out in a more leisurely fashion, as it did not interfere with the traffic on the lift.

Meanwhile at the Weaver Navigation Yard at Northwich, the various components which were to make up the superstructure for the electrical conversion, were being prepared in readiness for installation. To this end, a number of outside companies were employed by the Trustees to provide materials and equipment, as tabled below:

EQUIPMENT	MANUFACTURER
★ Rimming steel columns and tubes	Mannesmann Tube Company of London
★ Cast iron gear wheels and pinions	P. R. Jackson and Co. of Salford
★ Brass and gunmetal castings	Sykes and Harrison Ltd. of Manchester
★ Castings for pedestals and counterweights	Cleghorn and Wilkinson Ltd. of Northwich
★ Steel plates and angles	Shelton Iron, Steel and Coal Company of Stoke
★ Drainage pump and electric motors	Royce Limited of Trafford Park
★ Shafting and gearing	Hall and Pickles of Manchester
★ Cast steel clips and rings	Edgar Allen and Company, Sheffield
★ Electric motors	Mather and Platt Limited of Manchester
★ Main motors, tramway controllers etc.	British Electric Plant Company of Alloa
★ Smaller electrical fittings & power supply	Northwich Electrical Supply Company
★ Wire ropes	W. D. Houghton & Co. Ltd. of Warrington
★ Rubber seals	David Moseley and Sons Ltd of Manchester

The rimming steel columns and tubes for the 'A' frames were fabricated at Dusseldorf in Germany and imported by the Mannesmann Tube Company. They were designed in relatively short lengths, with bolted connections for ease of transport and assembly on site. During the process of fabrication, however, an error was identified which cost the Trust an additional £2,016 to put right. It would appear an incorrect strength parameter had been selected for the rimming steel, with the result that the columns and tubes were initially designed with walls which were too thin for the loads to be carried. Replacement columns and tubes had to be fabricated accordingly with consequent delays.

Much of the finishing work to the bought in materials and components, such as turning and boring of the gear wheels and pinions, were carried out by the staff at the Weaver Yard. Once ready for assembly, the components were transported the relatively short distance to Anderton. Soon after, the superstructure for the electrical conversion, which included the five pairs of 'A' frames and the upper machinery deck, was erected over the hydraulic lift while it was still in operation, and the drive machinery for the eastern caisson installed.

The third and final stoppage to traffic was for a period of 20 days between 16 April and 5 May 1908, during which time, amongst other things, the eastern caisson was modified, with the removal of the hydraulic ram head and installation of side beams for the connection of the wire ropes. The wire ropes and cast iron counterweights were installed and by 1 May the aqueduct was refilled and the converted caisson made ready for testing.

The tests were evidently successful because it is recorded in the *Anderton Boat Lift - 1908* report that:

'. . . so well had the things been arranged that the caisson moved without the slightest hitch when the power was switched on.'

Between May and July 1908, the traffic through the lift was conveyed by the eastern, electrically operated, caisson thereby enabling the conversion of the western caisson to be carried out without further stoppages. The western caisson was successfully brought into operation at the end of this 12 week period.

The converted lift was formally opened by the Chairman of the Trust, Sir Joseph Verdin, Bart, at a ceremony on 29 July 1908, in the presence of 40 invited guests. There is little doubt that the occasion was marked by a good deal of pomp, as suggested in a subsequent article in the *Northwich & Knutsford Guardian*:

"Invitations were issued to a number of representative men in the district, in the Potteries, and in Birmingham, as well as several engineers. Those who responded boarded Miss Weaver, *a well appointed vessel under the command of Captain Mills, near the offices at Northwich. Towed by the* Water Fly, *it was not long before Anderton Lift was reached. Here the company disembarked and entered a flat, which was drawn into one of the tanks (caissons), and in passing into it Sir Joseph Verdin, with a pair of scissors presented to him by Mr. Saner, expeditiously cut the ribbon spanning the entrance, at the same time saying "I declare this lift open. The work has been well and effectively done." The ascent was then made, and on reaching the top an explanation took place of the wonderful piece of mechanism. All were loud in their praises of the engineering skill displayed in the work of re-construction.'*

The article goes onto to record the formal proceedings at a celebratory lunch which followed at the Angel Hotel in Northwich. In the speeches after the repast, Colonel Dixon stated that:

'They had seen the lift, newly reconstructed by Mr. Saner, in a manner greatly to his credit. (Hear, hear.) They had noticed how beautifully it was worked and managed. They had also seen the Weaver and they could judge whether this undertaking was such as to secure their confidence. Sir Joseph Verdin had his hand upon everything, and with him at the head they might depend on it, success would attend the effort.'

In responding, Sir Joseph Verdin had the following to say about the lift and the Weaver Navigation:

'The Anderton Lift, which had been reorganised by their present engineer, Mr. Saner, was he thought not only a very lofty monument to his skill and ingenuity, but also likely to be a great factor in developing trade between the port of Weston Point and the county of Staffordshire, and it now only remained for the commercial genius of the county to utilise it to its fullest extent, and thus enable them to justify to themselves and others the wisdom of the large expenditure they had made.'

★ ★ ★ ★ ★ ★ ★ ★ ★ ★

The final cost of the conversion to electrical operation was £25,869, over £10,000 more than the estimate given to the Trustees just four years earlier. The greater part of the over-expenditure was explained by Saner to be due to the increasing cost of iron and steel. On the positive side, however, traffic passing through the converted lift soon grew beyond the pre-conversion peak and within five years the electrical lift was handling an annual tonnage of 225,000 tons (the largest annual tonnage achieved by the hydraulic lift was slightly over 192,000 tons in 1904). The conversion was heralded as a great success both technically and financially.

The money raised to enable the lift to be converted was gradually repaid over the years after the conversion of the lift but for a number of reasons, the loan was not finally settled until 1935.

So Saner had achieved what he had set out to do, albeit at a greater cost than originally envisaged. He had masterminded the conversion of the Anderton Boat Lift from hydraulic to electrical operation over a period of nominally 27 months, and as promised, with just three relatively short stoppages to the flow of traffic through the lift (Saner originally asserted that the conversion could be done with three stoppages of two weeks each, whereas in practice it was completed with the same number of stoppages amounting to nominally seven weeks). The nearest equivalent to this engineering feat by modern day standards would be widening the M25 motorway from three to four lanes whilst maintaining traffic flows. Whilst the converted lift cannot be acclaimed to be elegant in architectural terms, it certainly stands as testimony to Saner's engineering management skills.

★ ★ ★ ★ ★ ★ ★ ★ ★ ★

Similarly to Chapter 4, which deals with the original hydraulic lift, it would be wrong not to include a detailed description of the construction and operation of the electrical lift as it existed after 1908.

As is evident from the preceding text, the converted boat lift, as it was in 1908 and as it can generally be seen today, effectively consists of two structures in one. Many of the elements of the original 1875 structure, albeit in an altered form, were retained but these were overshadowed by the more dominating 1908 structure.

The canal basin was largely unaffected by the conversion works, although the nearest salt chute on the western side of the lift was dismantled to allow room for the support columns of the new structure. The wrought iron aqueduct also remained generally as it was in 1875 but was strengthened by the installation of additional cross girders under the flooring of the troughs.

Similarly, the basic elements of the original masonry lift basin were retained but with extensive concrete foundation works, as referred to above. These foundation works included the construction of a solid floor, which necessitated the excavation of 5 feet of soil from the bottom of the original basin and its replacement by concrete. The upper surface of the concrete slab was dished, being 5 feet 6 inches thick around the edges and 4 feet 6inches thick in the middle along the length of the basin. The floor slab underpinned the existing masonry side walls and was reinforced laterally by the inclusion of bulky steel 'H' joists at 4 feet centres. The approximate volume of excavation and, hence, concrete placed was 16,000 cubic feet. These foundation works, together with the provision of a second pair of river end gates, facilitated the conversion of the lift basin from a wet well to a dry well. This alteration gave three significant benefits to the new lift, as follows:

★ Reduced contact of the caissons with river water, and hence less risk of corrosion.

★ Improved access to the lift chamber for maintenance purposes.

★ Avoidance of the water displacement effects on

A view of the conversion works at the top of the lift, after the installation of the new 'A' frames along the eastern side. Behind them, the eastern caisson in its fully raised position can be seen; it is still, at this satge of the works, being operated hydraulically. It is not clear whether the 'A' frames on the western side are in place in this photograph.
Courtesy The Boat Museum, Ellesmere Port

With the inner row of cast iron gear wheels installed on the machinery platform over the eastern caisson, a section of the outer steel support beam of the platform is carefully manoeuvred into place.
Courtesy The Boat Museum, Ellesmere Port

This superb photograph shows the conversion works in an advanced stage, with all 10 of the 'A' frames, the machinery platform and most of the cast iron headgear in place. The dramatic increase in size over the old boat lift can be appreciated. Note the new end gates and the lack of a footbridge in this view, the old wooden bridge having been dismantled. New brick approach steps have been built ready for a replacement. Colin Lynch collection

A close-up of the front of the lift at river basin level, clearly showing the new river end gates, needed due to the adoption of a dry well system. The replacement footbridge is in the process of being floated into position. Unlike its wooden predecessor, this steel bridge crossed the lift entrances in a single span, with increased headroom. It was consequently a more substantial structure than the one it replaced. Once the barge is in the correct position, the new bridge will be lifted by the derricks onto the approach steps. Courtesy The Boat Museum, Ellesmere Port

A view of the eastern side of the lift, with the 'A' frames, machinery platform and some of the headgear for the eastern caisson in place. The upper part of the original hydraulic lift structure can be seen beneath the heavy machinery platform.

A view of the works on the machinery platform looking southwards, with the Winnington Works once again in the background on the far bank of the river. Work has progressed significantly since the previous picture, with the cast iron headgear for the eastern caisson appearing to be complete and the caisson possibly already in electrical operation. The installation of the head gear for the western caisson is progressing.

Both courtesy The Boat Museum, Ellesmere Port

The cast iron headgear and drive mechanism being installed over the western caisson. The bearings for the drive shaft can be seen in the foreground and the pedestals are lined up on the left, awaiting the installation of the gear wheels. The men appear to be working on the base plates for the motor and gearbox.

The drive mechanism for one of the caissons, with the electrical motor in the foreground on the right, the gear box and drive shaft on the left and the gearwheels in the background.

A close up of the completed drive mechanism for one of the caissons, with the motor and gearbox in the foreground. A wooden cover has been lifted off the motor and set down on its side behind. The motors, which were supplied by Mather & Platt of Manchester, appear to be relatively small given the weight of the caissons. It must be remembered, however, that each caisson was balanced by the counterweights which were hung from wire ropes down the sides of the lift and the motors were only needed to overcome frictional forces in the system.

All courtesy The Boat Museum, Ellesmere Port

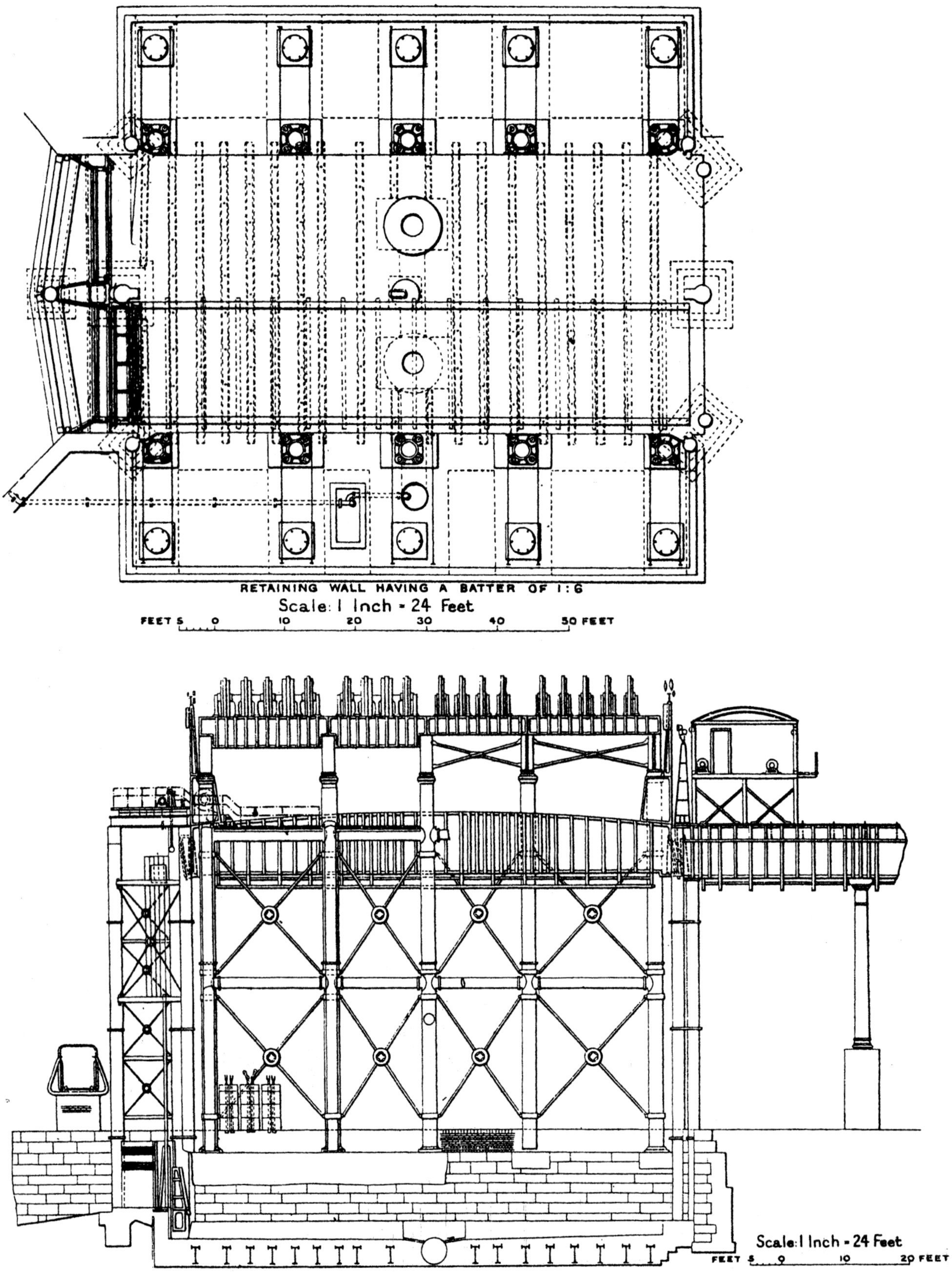

PLAN (TOP), SIDE (BOTTOM) AND FRONT (BOTTOM RIGHT) ELEVATIONS OF THE ANDERTON BOAT LIFT AFTER CONVERSION TO ELECTRICAL OPERATION

PLAN SHEWING FIXING ARRANGEMENT OF THE 30 H.P. ELECTRIC MOTORS

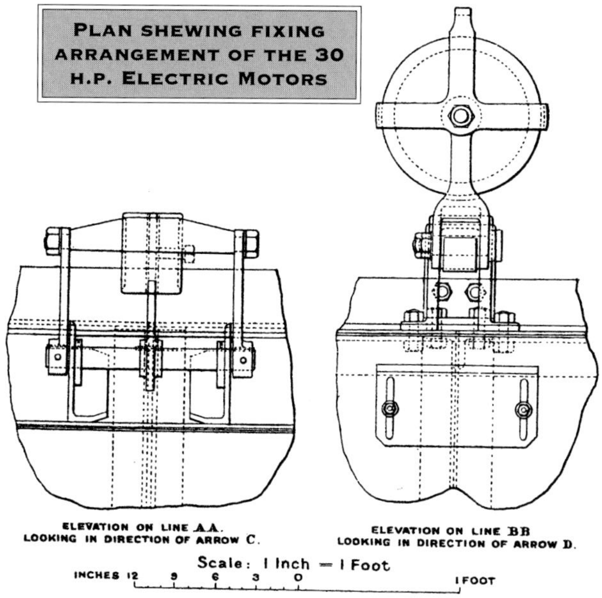

PLAN AND ELEVATIONS OF THE AUTOMATIC MECHANISMS WHICH WERE INSTALLED ON THE AQUEDUCT AND CAISSON GATES TO ENABLE THE PAIRS OF GATES TO BE RAISED AND LOWERED TOGETHER, WHILST ALSO ALLOWING THE CAISSONS TO DESCEND WITH THEIR GATES DETACHED FROM THE AQUEDUCT GATES

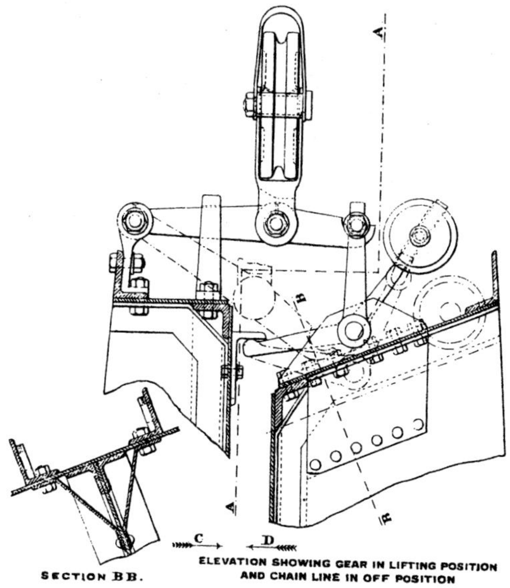

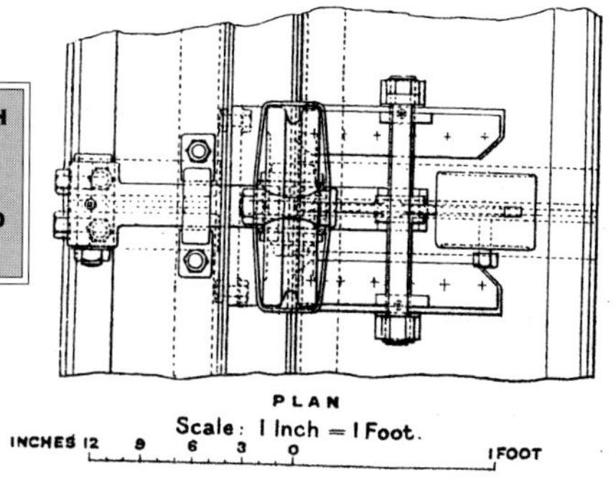

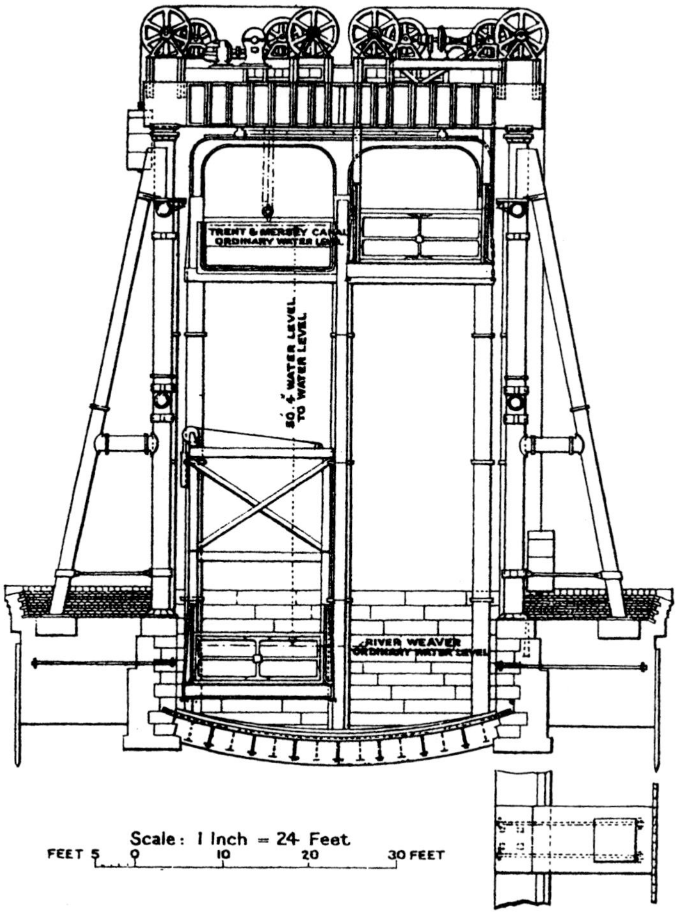

PLANS AND ELEVATION OF THE TOP OF THE 'A' FRAMES SHOWING THE BALL AND SOCKET CONNECTIONS WITH THE MAIN CROSS BEAMS OF THE MACHINERY PLATFORM

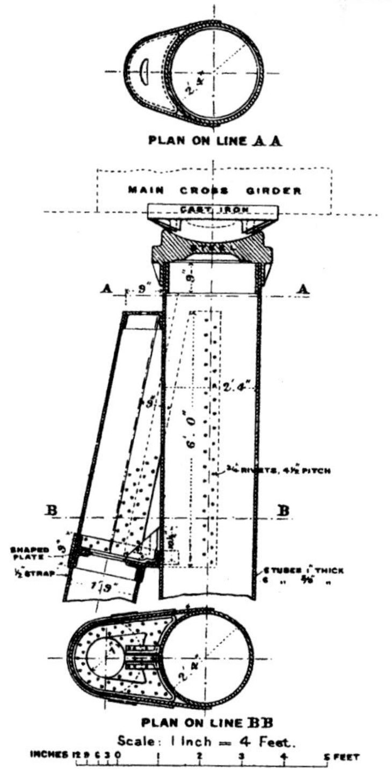

The newly converted electrical boat lift in all its glory. Its massive bulk straddles the retained elements of the hydraulic lift. In comparison to earlier photographs, note the changes including the sunken areas for the bases of the 'A' frames, wire ropes and counterweights at ground level on the western side, the new steel footbridge across the entracnes to the lift and the truncated salt chute alongside the aqueduct. Although not an elegant structure, Saner's achievement is having converted the lift to electrical operation with only minimal disruption to the traffic passing between the Weaver Navigation and the Trent & Mersey Canal.
Courtesy The Boat Museum, Ellesmere Port

lowering the caissons, hence resulting in much simplified operational procedures.

The cast iron columns and much of the wrought iron framework of the 1875 lift structure were retained. Similarly for the two wrought iron caissons, although these were modified to facilitate their being suspended from wire ropes as opposed to being supported from beneath by the hydraulic rams.

Not surprisingly, the most significant part of the conversion was the massive new iron and steel structure which straddled the smaller 1875 lift. The sole function of this construction was to support the heavy cast iron lifting gear, high above the lift basin, thereby enabling the caissons to be hung below on wire ropes. The machinery deck is nominally 60 feet above ground level and is supported by ten 'A' frames, five along each side of the lift basin. Each 'A' frame is of bolted steel construction and comprises a hollow tubular vertical column with an angled hollow tubular bracing column. At approximately mid-height, the vertical and diagonal columns are tied together by a horizontal tubular brace. Each 'A' frame is fixed at ground level by means of bolted connections into the strengthened masonry walls of the lift chamber and the new concrete counterfort walls. The whole structure is given longitudinal stability by means of hollow tubular struts at one third and two third height, together with small diameter diagonal cross struts between adjacent 'A' frames along the sides of the basin.

The machinery deck itself consists principally of multi-plated, riveted steel beams. Five heavy 'I' section cross beams are supported at each end by the vertical columns of the 'A' frames with special dish connectors which were designed to permit independent rotational movement of the beams. The five cross beams, in turn, support longitudinal beams, onto which the cast iron headgear and electric motors were mounted.

As was the case for the 1875 lift, each caisson in operational condition (ie. when filled with water) weighed nominally 252 tons. However, in the converted lift, each caisson was balanced by 18 cast iron counterweights, each weighing nominally 14 tons. The counterweights hung vertically down the outer sides of the lift structure and were connected to the caissons by wire ropes. To facilitate the required movement of the caissons and counterweights, the ground level on either side of the lift basin had to be lowered locally by nominally 4 feet. This necessitated the construction of brick retaining walls to support the exposed vertical faces. When the caissons were fully raised the counterweights were at the reduced ground level and when the caissons were fully lowered in the dry well the counterweights were close to the underside of the machinery deck.

The Angel Hotel in Northwich where the lunch to commemorate the successful conversion of the lift was held on 29 August 1908. The building is showing distinct signs of subsidence and was later demolished.
Colin Lynch collection

Each counterweight was connected to its respective caisson by two wire ropes; one passing over the pulley on the machinery deck immediately above and then vertically downwards to the outside edge of the caisson; the other passing over two pulleys and then vertically downwards before being connected to the inside edge of the caisson. As previously described, every second counterweight had two, additional, safety ropes which, instead of being connected to a caisson, were fastened to the pulley immediately overhead. Under normal operational conditions, these safety ropes contained some slack and, on raising of a caisson, would wind round the pulley with all the weight of the caisson and counterweights being taken by the working ropes. However, in the event of slippage of the working ropes over the pulleys, say due to

The Weaver Navigation Trustees survey boat Miss Weaver, *moored in front of the Company's offices at Navigation Road, Northwich. The vessel was used during the opening ceremony for the converted boat lift on 29 August 1908; it departed from here under tow of the tug* Water Fly *to the Anderton Basin, where the party of invited guests and dignataries aboard transferred onto a Weaver flat to ascend the lift.*

Mike Day collection

loss of weight of the caisson, the safety ropes would become taut and half the weight of the affected counterweights would be taken out of the system, thereby, hopefully, preventing further slippage (this, together with the inclusion of safety pawls, were the main modifications that came out of the review of Saner's proposals by Sir Benjamin Baker).

The lifting mechanism for each of the caissons was driven by a 30 horse power electric motor, which operated off a 440 volt supply at a fixed speed of 750 r.p.m. As stated in the *Anderton Boat Lift – 1908* report:

'This speed is reduced by means of an enclosed worm and wheel to 18 revolutions per minute. The worm wheel actuates a longitudinal shaft which conveys the power to two short cross shafts, making 12 revolutions per minute, placed at equal distance from the centre and from each end of the pulleys. By means of bevel wheels, the worm wheel actuates two longitudinal shafts running the whole length – 70ft. – and having pinions engaging with each pulley. The speed is by this gearing reduced to three revolutions per minute, and as the pinions are 12in. in diameter and the wheels 6ft., there is a further reduction, so that the main pulleys only make one revolution in two minutes. Every pulley is therefore directly geared to the shafts, and as all the teeth are helical, there is an absence of noise and backlash, the humming of the electric motor only being noticeable. The placing of the cross shafts also ensures the simultaneous movement of the pulleys by equalising the torsion on the main shafting.'

With the rate of turn of the main pulleys of 0.5 r.p.m and an outer diameter of 6 feet (1.82 metres), the resulting rate of lift of the caissons would have been nominally 9 feet per minute (1.2 metres per minute). With the time taken to close and open the gates, the overall lift from the river level to the canal level took around 10 minutes. Due to the gross weight of the relevant counterweights being the same as the weight of the filled caisson, the only power required from the main electric motors was that sufficient to overcome the friction of the pulleys on their bearings, the losses within the gearings and other minor losses. On this basis, the 30 horse power were over-sized, the additional capacity having been provided to allow for the possibility of weight imbalance due to varying depths of water in the caissons resulting from changes in river level conditions.

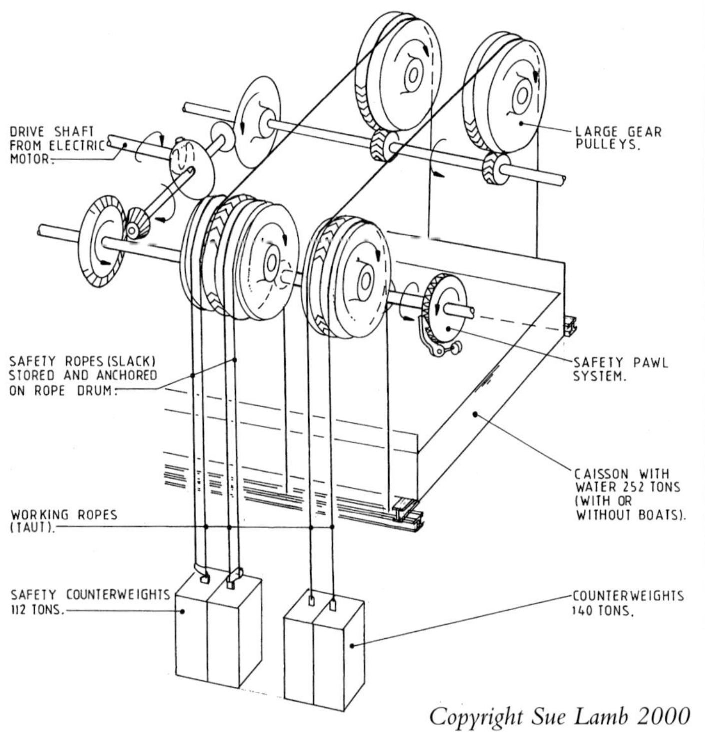

Copyright Sue Lamb 2000

DIAGRAM OF OPERATING SYSTEM OF THE CONVERTED LIFT, SHOWING THE CAISSON, LIFTING ROPES, SAFETY ROPES, HEADGEAR AND COUNTERWEIGHTS.

In practice, the need for this over capacity was very much negated by the use of adjustable sealing frames. This innovative idea was adopted to enable the lift to be functional at most states of the River Weaver. Adjustable frames were installed at the river end of the dry well, against which the caissons sealed at the end of their descent. These frames could be raised or lowered, according to the state of the river, so that the water level in the caissons better matched the level in the river, thereby preventing too much or too little water being in the caissons prior to lifting.

Rather surprisingly, the operation of the new electrical lift was found to require nominally half the power used by the hydraulic lift with the rams being powered by electric pumps (circa 1902).

Another innovation was the devices installed between the gates at the lift end of the aqueduct which enabled each aqueduct gate and it's corresponding caisson gate to be raised and lowered by means of a single mechanism. Hitherto, it had been necessary to have separate mechanisms for each gate. With the new arrangement, on raising a caisson, the caisson gate would automatically engage with the lifting gear of the aqueduct gate, thereby enabling both gates to be raised together to permit the passage of boats. Prior to descent, the gates would be lowered together and the device would automatically disengage, enabling the subsequent, unhindered, lowering of the caisson.

As a consequence of the 1908 conversion, the entire operation of the boat lift and the associated gates could be carried out in the control house above the aqueduct by means of electric tram-type controllers. This arrangement had the benefit of reducing the lift operator's workload. As explained in the *Anderton Boat Lift – 1908* report:

'. . . formerly, he, (the operator in the control house), was obliged to remain at the valves all the time the caisson was at a joint, either easing the pressure or applying more as either the accumulator or the caisson pressure became the greater. Now, as soon as the power is switched off everything is at a standstill, and nothing moves until it is again applied.'

The control cabin is colourfully described by Hugh

McKnight in *The Shell Book of Inland Waterways* in the following words:

'A keeper is in charge of the control cabin, a place of brass dials and levers like an old-fashioned railway signal-box.'

In addition, the conversion allowed a reduction in the workforce employed on the operation of the lift. Since 1896, when electrical power was first used to operate the gates, the lift required three men full-time, with a fourth man in part-time attendance. With the new set-up, only two men were needed on a full-time basis, one at the upper (control house and machinery deck) level and the other at ground level. A third man was deemed necessary, however, to act as cover in the event of illness. He was otherwise employed assisting boats in and out of the caissons and maintaining the machinery.

* * * * * * * * * * *

To close the chapter, here is another extract from the article in the *Northwich & Knutsford Guardian,* which records a speech to the assembled guests by Mr. Sandeman, who had been Engineer to the Trust from 1872 to 1877:

"The designer of the reconstructed Anderton Lift". He had much pleasure in proposing the health of Mr. Saner, who reconstructed the interesting lift they had seen that day. He thought the efficiency with which it performed its functions spoke highly of the genius displayed. It worked more economically and with greater effect than its predecessor. It worked so satisfactorily that it reflected the greatest credit upon him. As they knew the old hydraulic hoist was commenced by Mr. Leader Williams, and it devolved upon the speaker and another to put it into execution. After fulfilling its purpose it had now been replaced by the modern appliance. Work extending over 33 years was not a short life for a machine of that kind, considering the rapid progress in engineering science of recent years. Doubtless this was due to the care Mr. Saner and Mr. Wells had taken of it. Mechanical appliances which were thought very wonderful even 30 years ago had become quite antiquated, and as the Americans termed it had to be scrapped. The use of electricity as economical means of conveying power was gradually dominating all other systems. There was a great personal satisfaction to an engineer in the accomplishment of a successful piece of work, irrespective of the approval of others, and he had no doubt that Mr. Saner experienced that satisfaction. (Applause)'

The cast iron plaque on the north east guide column of the boat lift which records the design and construction of the hydraulic lift in 1875, and its subsequent conversion to electrical operation and re-opening in July 1908. Author

INFORMATION SOURCES FOR CHAPTER 7

Minutes of the Proceedings of the Weaver Navigation Trustees - 1904 to 1948. *British Waterways' Regional Office, Northwich.*
Anderton Boat Lift – 1908. *Paper by J. A. Saner to the Weaver Navigation Trustees.* The Engineer, *14 July 1908.*
Article in the Northwich & Knutsford Guardian, *29 July 1908.*
The Anderton Boat Lift - An Examination of its Impact on Local Trade and Population, and of the Extent to which Trading Figures and Working Conditions Reflect National Economic Trends. *Dissertation by Sharon Rosalind Murray, 1990, Chester College of Higher Education.*
Miscellaneous Correspondence File Records of the Weaver Trustees. *Ellesmere Port Boat Museum.*
The Shell Book of Inland Waterways *by Hugh MacKnight. David & Charles, 1975.*

An atmospheric view southwards over aqueduct and converted boat lift, with the Winnington Works belching smoke in the background. Despite the sunshine, the shards of ice in the canal basin indicate that it is wintertime. Courtesy The National Waterways Museum, Gloucester

CHAPTER 8
OPERATION OF THE ELECTRICAL LIFT (PART 1)

'The Anderton Lift continues to work with perfect satisfaction . . .'
Extract from J. A. Saner's *Annual Report to the Weaver Trustees* on 11 November 1909

By all accounts, the conversion of the Anderton Boat Lift from hydraulic to electrical operation, as masterminded by Col. J.A. Saner, was a resounding success.

The prognosis was good even in the early days of the converted lift. For example, part way through the conversion process, when the lift was under single caisson operation between May and July 1908, the electrified No.1 (Up River) Caisson made 2,819 movements, with a total of 3,671 boats being transported, (ie. nominally 50 boats a day). After the completion of conversion, between the end of July and the end of October 1908, each caisson had made 2,987 movements with 3,562 boats being transported. As a result of these figures, the Trustees were advised that one of the caissons on its own would have been capable of handling the whole of the traffic in existence at that time.

Just a few years after the conversion of the lift Saner wrote a special report to the Trustees on the working of the lift. In the report, which was dated 19 October 1910, Saner advised that:

'The No.1 (or Up River) Caisson was opened for traffic on May 5th, 1908, and up to the 8th instant has made 19,199 strokes either up or down, passing 22,866 boats.

The No.2 (or Down River), Caisson was opened for traffic on July 20th , 1908, and up to the 8th instant has made 11,905 strokes either up or down, passing 14,084 boats.

This makes a total of 31,104 strokes and 36,950 boats . . . . The total tonnage passed has been 446,796 tons or an average of 14.36 tons per stroke.'

Records reveal that as early as 1909 the total tonnage transported by the lift had exceeded the pre-conversion peak of 192,181 tons, which occurred in 1906. Furthermore, as previously stated, within five years the annual tonnage had risen to 225,000 tons.

In the report, Saner comments on the operation of the electrified lift in the following way:

'So far there has been no trouble, (with the working of the converted lift); on one occasion the safety pawls were accidentally put into action when the No.2 Caisson was moving, but in spite of the fact that 500 tons was brought to a sudden stand nothing serious occured, and the slight damage done was repaired in a few hours. There is no doubt the resilience of the ropes eased the shock, which would otherwise have been destructive.

None of the gearing has had to be touched except the discs carrying the first-motion worm wheels, which were found to be unsound castings and were replaced by the makers without charge.

The electrical machinery has given no trouble so far; a slight exception has been in the solenoid coils of the magnetic brakes, which are somewhat exposed to the action of grease; they have now been covered up with wooden boxes.

The bearings, wheels and ropes shew no signs of wear so far; with care it should be many years before the gearing requires anything doing to it, and although it may be desirable to make a special examination of the ropes in say two years I do not anticipate they will require renewal for at least ten years from being new.

. . . After the experience of the last two and a half years I am quite satisfied with the work, and do not see why, excluding accidents, but including a sufficiency of proper paint and oil, it should not last indefinitely, . . .'

Saner was obviously attempting to be as positive as possible regarding the operation of the converted lift, whilst at the same time advising of the various problems which had been encountered to date; a rather difficult balancing act! Under the circumstances, however, it is not particularly surprising that Saner, the architect of the converted lift, gave a report to the Trustees which made light of any early difficulties.

Saner was also able to report that, for 1909, the working expenses of the lift were £492, which compared favourably with those prior to conversion in 1907 of £650. The direct income to the Trust due to the working of the lift was £506 in 1907 and £578 in 1910. It should be remembered, however, that rates being charged at that time for use of the lift were set at a relatively low level, the main benefit being derived from the increased usage and hence higher income from the Weaver Navigation as a whole.

Saner took the opportunity in his report to warn the Trustees of the need for relatively frequent and costly

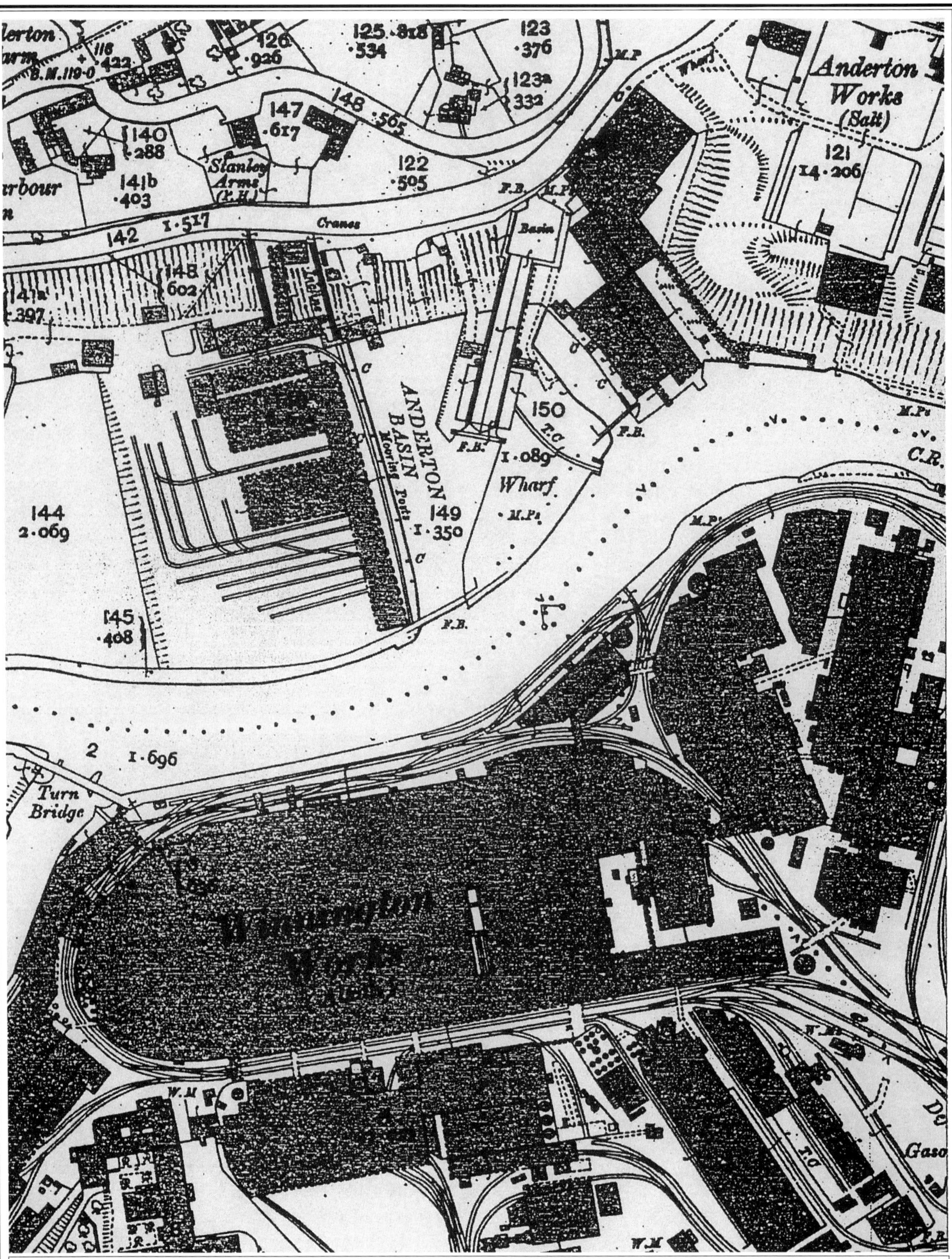

The 3rd edition (1910) 25 inch OS for the Anderton – Winnington area, showing the changes to the basin following the conversion of the lift. Courtesy Cheshire Record Office, Chester. Crown copyright reserved

replacement of the rubber seals on the various joints and gates in the lift. In addition, he reminded them of the highly corrosive atmospheric conditions in the Anderton area and the consequential need for frequent and regular expenditure on surface treatment:

'Experience has proved that ordinary paints are of no use in that particular neighbourhood; all the tubes and girders, which are well painted with 3 and sometimes 4 coats two years ago, are peeling and shew rust, and I am, as previously reported, now using a black enamel made from bitumen, which is found to

stand somewhat longer than the paint. The important thing in all steel structures is to coat them regularly, and I think it will be necessary to do the Lift without fail every third year, and perhaps portions more frequently. This does not apply (just) to the Lift but to any steel work in the neighbourhood. Owing to this corrosion it may be necessary from time to time to renew portions of the bottoms of the old part of the aqueduct, but the main girders appear to be in good serviceable order .... It is early to say what the cost of repairs will be, but I do not anticipate it will be nearly so large as formerly.'

There is no clear record of whether Saner's recommendations regarding painting were adhered to. It can be assumed, however, that in the early years, whilst he remained as Engineer to the Trust, they were. In later years however, when it could be argued that the paint systems were better, painting on the lift appears to have been done, at best, once every 4 or 5 years.

★ ★ ★ ★ ★ ★ ★ ★ ★ ★

During subsequent years, the operation and management of the lift were seemingly relatively trouble free, but before long a problem arose with respect to the condition of the wire lifting ropes. By 6 April 1914, the situation had become severe enough for Saner to present a second special report to the Trustees on the working of the lift. Having carried out a detailed inspection of the ropes on the No.1 (Up River) Caisson, he was able to report the following facts with regard to the longer lifting ropes, which pass over two pulleys to connect to the inside face of the caisson (it should be noted that each of the lifting ropes was $1^1/_4$ inches in diameter with six strands, each strand having 19 wires):

'In the longer ropes:
6 are apparently perfect
1 has 1 wire broken
1 .. 2 wires broken
3 have 3 wires broken
2 .. 4 .. ..
2 .. 5 .. ..
1 has 6 .. ..
1 .. 9 .. ..
1 .. 13 .. .. (Tested)
All the ropes however naturally shew some signs of wear.'

Perhaps surprisingly, the shorter ropes that connect to the outside edge of the caisson, and hence pass over just one pulley, were found to be in reasonably good condition.

On liaison with the makers of the ropes, the one with 13 wires broken was sent away to Sheffield Testing Works Ltd. for independent load testing. The findings were that the intact sections of the rope not required to pass over the pulleys, which consequently were never bent in operation, withstood more than the guaranteed load. Furthermore, the section with the broken wires still provided an adequate safety factor of 7 to 1. It was concluded that the main cause of the wires breaking was fatigue, arising from the regular bending and straightening of the ropes as they passed over the pulleys. It was surmised that the longer ropes were in poorer condition because

APPENDIX III.

The Sheffield Testing Works, Ltd.,
Blonk Street,
SHEFFIELD, *March 17th, 1914.*

Report No. 37533.
The following are results of Tensile Tests of two samples cut from a Wire Rope.
Received 9th March, 1914, from J. A. SANER, Esqr., M.Inst.C.E., Weaver Navigation Trustees, Northwich.

Test No.	Marks	Described as.	Circumference of Rope.	Weight per Fathom	Strands. No. of Strands.	Strands. No. of Wire	Strands. Dia. of Wire inc.	Total No. of wires	Hemp Core.	Stress in Tons. Extension of rope inches on 50 inches. 1	5	10	15	20	25	30	50	MAXIMUM STRESS. tons.	Total extension of Rope on 50 inches.	Remarks.
			inches.				ins.													
Q643	—	Sample cut from wire rope ... (sound portion)	3·90	—	6	12 6 1 Soft	·084 084 ·086	114	Main	Zero	·08	·13	·20	·27	·33	·38	·71	59·90	1·32	3 Strands broke together, clear of the fastenings
										1	5	10	15	20	25	30	40			
Q643a	—	Sample cut from wire rope ... (portion containing broken wires)	3·73	13·9	6	12 6 1 Soft	·084 ·084 ·086	114	Main	Zero	·06	·15	·23	·30	.37	·45	·52	47·90	.70	4 Strands broke together, clear of the fastenings

SKETCH SHOWING POSITIONS FROM WHICH TEST PIECES WERE TAKEN.

J. A. SANER, Esqre., M.Inst.C.E.,
NORTHWICH.

(Signed)
FRANCIS J. SMITH.

A table of results of tensile tests on the wire lifting ropes, dated March 1914, contained within a special report from Col. J. A Saner to the Trustees.
Steve Astles collection

they passed over two pulleys, and therefore went through the bending and straightening process twice as often as the shorter ropes. The special report was concluded by Saner as follows:

'The ropes have now been greased over again and a new rope put in to replace the one removed. The rope with 9 broken wires will of course be watched and probably next year No.2 Caisson ropes will be examined, but at present they do not show any signs of wear so far as can be seen with the grease on.

It is, I think, evident that the 10 years life mentioned in my report of October, 1910, will be accomplished by the majority of the ropes, more especially the shorter ones . . .'

Four years later, the minutes of the Trust for 20 June 1918 record Saner advising that 12 new long ropes had recently been fitted to the No.1 Caisson, to replace those which had been in use since the 1908 conversion. Thereafter, ropes had to be replaced on a relatively frequent basis.

★ ★ ★ ★ ★ ★ ★ ★ ★ ★

By all accounts, the inter-war period saw business very much as usual for the boat lift, with continued operation and regular maintenance, particularly with respect to the ever-present problem of corrosion. Occasional replacement of worn out or damaged components was also necessary from time to time. In 1920, the ironwork of the original lift was recorded as being in very good condition but, rather surprisingly, some of the steelwork installed in 1908 was corroded and in need of urgent repairs. In the following year, the interior of the aqueduct and caissons were inspected and also found to be in good condition, whilst the exposed surfaces were suffering badly from corrosion.

Extensive repairs were undertaken to the aqueduct and the caissons in 1922, principally, one assumes, to deal with the ongoing problem of corrosion. The repairs carried out at that time also included the replacement of the timber planking to the walkways on both sides of the aqueduct. On completion, the aqueduct and caissons, together with many other elements of the boat lift, were scraped and cleaned before being re-painted.

The following year, a replacement electrical pump was installed for draining water from the dry well, the original pump being refurbished and retained as a spare and in 1926 the original wrought iron troughs for the collection of water leaking from the caisson and aqueduct gates had to be replaced due to corrosion.

★ ★ ★ ★ ★ ★ ★ ★ ★ ★

In the minutes of the Trustees on 10 November 1927, Saner advised of an accident very similar to one that had occurred seventeen years earlier:

'On the 7th instant, when the No.2 Caisson had descended

The unpowered narrowboat Euston *enters the electrical lift in 1931, with numerous other working narrowboats moored in the foreground. The narrowboat* Nora *has her stern up against the entrance gate to the western caisson and the packet* Weaver Belle *can be seen moored outside her dumb barge* Gowanburn, *behind in the basin.*

Courtesy The Boat Museum, Ellesmere Port

Also taken in 1931 was this view of the interior of the control cabin, with the Lift Attendant operating the controls for the western caisson. The controls on the left were for the eastern caisson, including the lever in the foreground which operated the pawl system; a similar lever for the western caisson was located on the other side of the metal dividing screen. Courtesy The Boat Museum, Ellesmere Port

and the No. 1 Caisson was ascending, the operator in charge pulled the wrong lever by mistake and applied the 'pawls' to the ascending caisson instead of the other. Fortunately the resultant damage was not very serious. There was no delay to traffic, which is using No. 2 Caisson during the repairs; the latter should be complete in the course of a few days.'

It should be noted that the pawls were safety devices, similar to those still in use on lock gates to hold the paddle sluices in the open position. The pawls interlocked with ratchets on the cast iron headgear and were intended for use either when the caissons were being lowered or when static at the river level, to avoid the possibility of the motors unintentionally being driven backwards by out-of-balance forces, perhaps caused by unexpected loss of water from the caisson. The pawls, which were operated from within the control building, had to be disengaged during raising of the caissons. So accidental engagement of the pawls on an ascending caisson would bring it to a sudden and immediate halt. As suggested by Saner in relation to the previous incident, the only feature that might have prevented serious damage and perhaps injury was the inherent elasticity of the wire ropes.

★★★★★★★★★★

In January 1928, severe weather conditions resulted in a loss of power to the boat lift for 30 hours, which in turn led to a significant build up of traffic on both sides of the lift. Possibly for the first time ever, the lift was worked on a Sunday (8 January) in order to clear the backlog. In the same year, extensive repair works were undertaken which included the replacement of rubber seals and the de-scaling, by chemical means, of the discharge pipe from the pump which drained the dry well. The construction of new gates for No.1 Caisson was also put hand during 1928, with their installation the following year. The old gates were badly corroded and had been damaged by the frequent impact of boats. Minor repairs were carried out at the same time on the armature of the main motor for No.2 Caisson.

The following year, repairs and replacements on the lift were continued. This included the replacement of the worm gearing to the river gate on No.1 Caisson, which was badly worn whilst, once again, the inside of the caisson was scraped and re-painted. The footbridge across the river

A commercial postcard view of the Anderton Basin from about 1930, with the boat lift in the background. The footbridge in the foreground carried the Weaver Navigation towpath across the entrance to the basin. The photograph may well have been taken on a Sunday as there is little evidence of any activity. Courtesy The National Waterways Museum, Gloucester

entrance to the lift, which had originally been installed some 20 years earlier, was found to be badly corroded and was consequently dismantled and taken to the Weaver Navigation Yard at Northwich for extensive refurbishment. The bridge was eventually reinstated in May 1930.

New gates were installed in No.2 Caisson in 1930 and, at the same time, essential repairs were carried out to the gate lifting equipment. The opportunity was then taken to scrape and re-paint the internal surfaces of the caisson. Not for the first time it was recorded in the minutes that:

'The electrical plant as well as steel and ironwork of the Lift suffers from the local atmospheric conditions.'

Major re-painting of the structure was commenced during the following year, this time with specialist tar and rubber solution supplied from the nearby Brunner Mond works. Other essential repair and replacement works were carried out, including the replacement of three of the wire safety ropes. The painting works continued through 1932, whilst work was started at the Northwich Yard on the fabrication of replacement gates for the river end of the aqueduct. These were eventually installed two years later.

The same year, new hours of working for the lift were put in place by agreement with the canal carriers who paid the tolls. The new working times were 7 am. to 6 pm. during the months of October, November and March; 8 am. to 6 pm. during December, January and February; and 6 am. to 10 pm. for the months April to September inclusive. The summer working hours would surely have necessitated double shift working whilst single shift working would suffice at other times of the year. These new hours were estimated to give a saving to the Trust of approximately £150 per year.

Although it is not made clear in the records, it is probable that, by this time, the lift was manned by a Lift Attendant and at least one, if not two, assistants, and possibly still a clerk to collect any tolls and keep records of tonnages carried. Traffic through the lift, principally in relation to the Potteries and the salt fields, was still sufficient to warrant double caisson working. There was consequently little spare time in the working week for the Lift Attendants and their assistants to carry out maintenance duties, some of which, therefore would have been done at weekends when the lift was closed to traffic. Other, more arduous tasks, such as painting of the structure and rope replacement would also have been done at weekends.

The Lift Attendant's main duties, at that time, would have been carried out within the control building and would have included operating the aqueduct and caisson gates, raising and lowering the caissons and application of the safety pawls. He may also have carried out other tasks at the canal level, such as operating the penstocks to drain the chamber between the aqueduct and caisson gates prior to lowering the caissons. One of his assistants would have worked at ground level, operating the river and caisson gates and assisting boats in and out of the lift. There may also have been another assistant working at canal level to collect the tolls and assist boats in and out of the caissons from the aqueduct.

When not busy managing traffic, the Lift Attendant and his assistants would carry out routine maintenance tasks,

Mersey Weaver & Ship Canal Carrying Company narrowboats in the top basin in 1935. The butties Aboukir *and* Punch *can be identified, whilst boatwomen and children await the return of their menfolk with orders. These Stoke-registered boats are Weaver-bound to collect pottery making materials. Note the centre walkboards, so crew members could reach the bow without treading on the cargo and for sheeting up. The upward curve of the tiller was an indication boats were 'stopped' and it also made cabin access safer and easier. Note the mops on the cabin tops which have been put to good use on the spruced up boats.* Courtesy The National Waterways Museum, Gloucester

Having just ascended the lift, a laden Mersey Weaver craft exits the eastern caisson and glides under the control cabin in the mid 1930s. The vessel is probably bound for the Potteries with a cargo of flints. As also the boats in the previous view, note the traditional painted decorations of flowers and castles. Courtesy The Boat Museum, Ellesmere Port

such as greasing the ropes, lubricating the operating mechanisms and generally keeping the structure tidy. Another of their duties would be the clearance of ice from the overhead cogs and pulleys that built up on very frosty days. At this time, it is very probable that the lift staff kept a rigorous look-out for any developing problems on the lift structure, such as localised corrosion, and would call in assistance, as and when necessary, so that they could be put right without too much difficulty. Painting of the structure and the heavier maintenance or repair works were generally left to fitters from the Northwich Yard.

★ ★ ★ ★ ★ ★ ★ ★ ★ ★

An amusing incident recalled by Jack Horton, who started working for the Weaver Navigation around 1932, took place around this time:

'One thing that did happen was, and we laughed about it [afterwards] *because they should have known better,* [was that] *in one period the boats went down* [the lift] *and they were discharged physically, the narrowboats, into bigger steamers which worked from Anderton Basin. There were two or three steamers that worked from Anderton Basin only, taking salt from Anderton to Liverpool or to Manchester which came down the lift. Now, somebody had a good idea, 'Why not lower the lift half way down and get your lumpers, as they called the men that shovelled, to feed down into the boat in the basin alongside the lift, put a chute in?' What they forgot was that the lift was held in suspension at this point* [and] *when they started to lump, discharge salt, the lift it was light and started to go up. They should have known that.'*

Presumably the safety pawl had been applied at the time of the experiment and the reason the caisson had moved upwards was due to the wire ropes slipping over the pulleys. Fortunately, the potentially severe consequences appear to have been averted.

★ ★ ★ ★ ★ ★ ★ ★ ★ ★

In July 1934, Saner, Engineer to the Trust for nigh on fifty years, since December 1887, finally retired. According to Jack Horton, Saner had been priming his son to take over from him as Engineer on his retirement but his ambitions were thwarted by the eventual appointment, by the Trustees, of his assistant, C. M. Marsh, as Engineer and Acting Manager.

The potential implications for the boat lift of the retirement of Saner should not, however, be overlooked. Saner was the designer and constructor of the 1908 electrical lift and would therefore have had a personal interest in ensuring its success. As Engineer to the Trust, Saner was in an influential position to ensure that the ageing structure was given as much attention as it required.

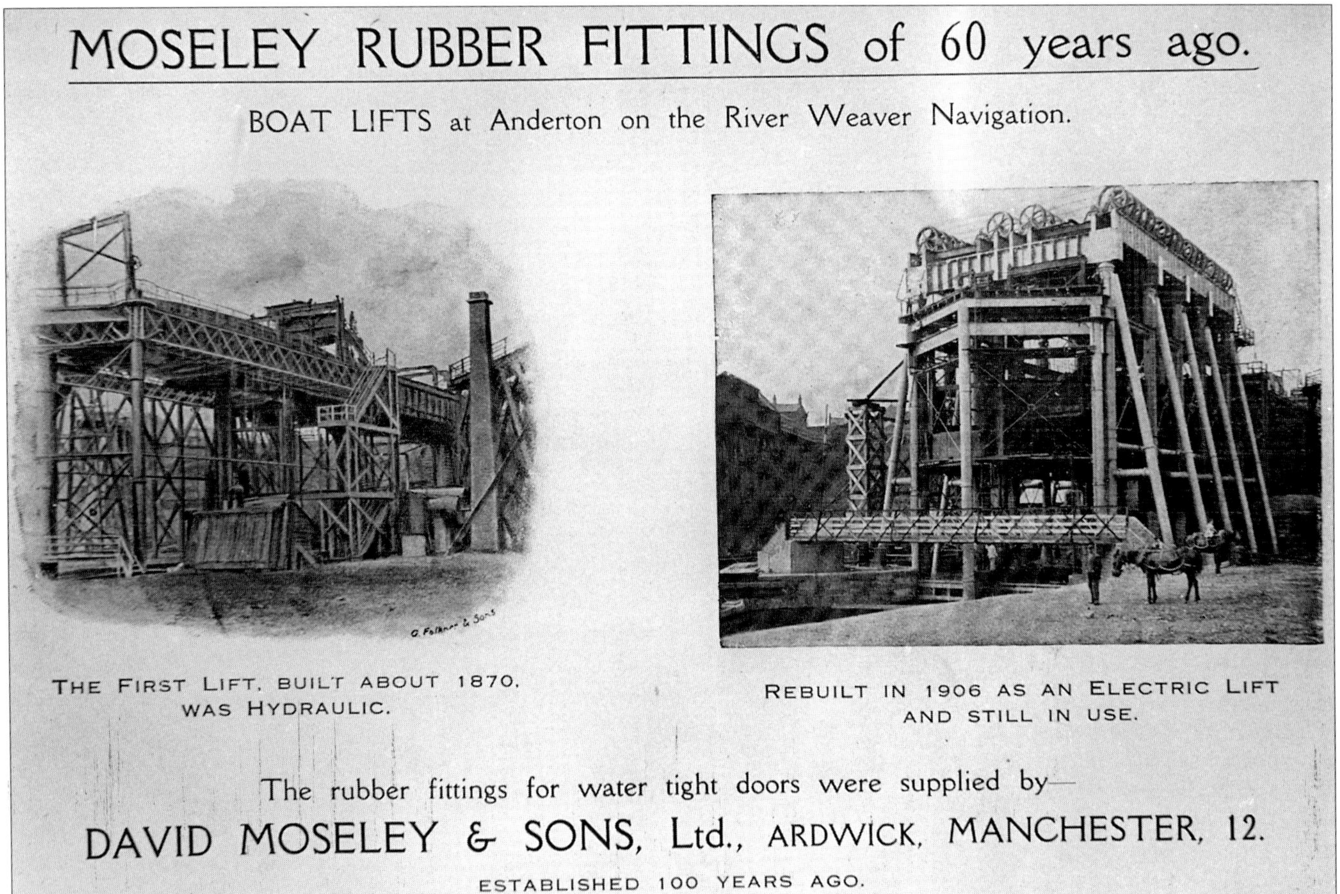

An advertisement for Moseley Rubber Fittings, circa 1930. The company had supplied fittings for making the doors water tight for both the hydraulic and electric lifts. Note the date given for the conversion; the year work commenced, rather than the year of completion, presumably because the new river doors for which Moseley's supplied fittings were constructed in 1906. Courtesy The Boat Museum, Ellesmere Port

Maintenance work on the drive equipment on the machinery deck, circa 1910. Greasing of the gear wheels and wire ropes was a regular duty.

Courtesy The Boat Museum, Ellesmere Port

Thereafter perhaps, without the advocacy of Saner, the lift gradually came to be regarded as just another structure on the Weaver Navigation.

★ ★ ★ ★ ★ ★ ★ ★ ★ ★

The ensuing years up to the onset of the Second World War saw ongoing repairs and renewals to the lift, including rope replacements and re-painting. Repairs were undertaken to many of the transverse floor girders on the aqueduct to counter the effects of corrosion. An extract from Marsh's report to the Trustees on August 1938 provides some interesting details:

'(a) PAINTING - The contract work was satisfactorily completed on August 10th: 850 gallons of tar and rubber solution have been applied to the structure. The total cost of the contract, supervision and materials has been £701. The Trustees' men are now employed painting the cabins and timberwork in stairways, etc.

(b) REPAIRS - The repairs to the aqueduct girders were completed on August 13th. 29 out of the 32 transverse girders have been repaired and 21 intermediate longitudinal girders have been renewed. The repairs have cost £461.'

Around this period, general painting of the numerous structures on the Weaver Navigation would have been done by a team of six or so men. Arthur Dodd, who in 1938 at the age of nineteen had been taken on as a labourer, advised that:

'The painters would start at one end of the river and go right through the whole system and when they finished they'd start again. They, (the structures), all took their turn unless there was some kind of disaster or any particular problems.'

This arrangement for painting the lift was confirmed by Brian Haskins, who was Area Engineer for British Waterways between 1968 and 1988. Brian suggested that by this means each structure on the waterway was painted once every four or five years.

★ ★ ★ ★ ★ ★ ★ ★ ★ ★

Whilst working as a labourer on the Weaver Navigation, Arthur Dodd was often employed carrying out menial tasks on the boat lift such as patch painting by means of a ladder and assisting with rope replacement. In September 1938 he suffered a serious accident whilst helping to haul a boat up a ramp at the Northwich Yard, which prevented him from working for six weeks. On returning to work he was assigned to light duties, which included stocktaking at Northwich. During this, Arthur Dodd recalls having seen several original models of engineering structures, including one of the converted boat lift. It transpires that Mr. Cawley, a Weaver Navigation employee, built the model of the electrical lift in 1913, presumably for Saner. It is understood that, at some stage, this model was taken to Milan in Italy for exhibition purposes. It is currently on display at the Canal Museum at Stoke Bruerne.

★ ★ ★ ★ ★ ★ ★ ★ ★ ★

At the start of the war, members of the Local Defence Volunteer Force, (the Home Guard), were detailed to keep the boat lift under observation during night time hours, as it was perceived as a potential target for German attack. The lower grades of staff on the Weaver Navigation, including Arthur Dodd who was a labourer at the time,

The model of the electrical lift which was built built by Mr. Cawley at Northwich in 1913. It contains a single caisson and measures nominally 7 feet long by 3 feet high by 2 feet wide.
Courtesey The Canal Museum, Stoke Bruerne

An excellent interior shot showing two working narrowboats in the east caisson; date unknown but probably mid 1950s. Both craft appear to be British Waterways 'Admiralty' class boats, which were built in the early 1950s with these unusual hooped tarpaulin supports, with walkboards along the top. Colin Lynch collection

were called up for active service and, for the war years at least, the waterway would have been operated with noticeably fewer men. Single men called up for service were paid by the Weaver Trust during their absence at 7 shillings a week and married men over the age of 21 at 10 shillings a week. The positions held by other members of staff, including the Lift Attendants, were designated as reserved occupations. (Arthur Dodd returned to work with the Weaver Navigation in 1945 having survived over three years as a prisoner of war).

In February 1940, severe frost caused damage to the masonry retaining wall of the canal basin, which resulted in leaks. Localised repairs were presumably carried out to the basin by way of repointing, and perhaps refacing, of the stonework and re-claying of the bed.

Records indicate that during 1941 and 1942 the original hydraulic rams were removed from the shafts under the dry well where they had been abandoned since 1908. It is likely that the redundant ironwork was needed for the war effort. This task would have entailed no small effort and could well have necessitated the closure of the lift for several weeks.

Around the same time, problems were encountered with one of the main lift motors, which had to be sent away for repair, most probably with the consequence that a relatively long period of single caisson working was necessary. Repairs were also carried out to the stonework on the river basin, to the pump discharge pipe and to the timber fenders, whilst three more of the ropes were replaced.

The rubber strips to the adjustable sealing frames (for the caissons in their lowered positions) were replaced in 1943 and the canal basin was dredged by the London, Midland & Scottish Railway (owners of the Trent & Mersey Canal since the Railway Grouping of 1923), the following year.

A serious incident took place in 1944, being reported to the Trustees by Marsh, as follows:

'LIFT: No.1 CAISSON - On April 21st an unfortunate accident occurred whereby the water drained off from the caisson. This was raised by the counter-balanced weights operating through the overhead gearing, and as a result the motor raced and damaged the armature winding, requiring complete overhaul at estimated cost of £150. The Trustees' Insurance Co. accept cost of renewal under their break-down policy.'

Unfortunately, the report given by Marsh is not sufficiently detailed to properly identify the cause and consequences of the incident. Perhaps one of the caisson gates had not been fully closed and the loss of water resulted in the main motor being driven by the imbalance of forces between the caisson and the counterweights. The report also fails to identify if there was any slippage of the ropes over the pulleys and whether there was any structural damage when the caisson eventually came to rest. It should, perhaps, be assumed that the caisson was in the process of being raised at the time, otherwise the safety pawls, on the basis that they had been applied correctly, would have prevented the incident.

The repairs to the damaged motor were carried out by British Electrical Repairs Ltd. and the motor was re-installed six months later. Again, single caisson working would have been necessary. The opportunity was taken to replace several of the ropes on the inoperable caisson. It is interesting to note that similar incidents are reported to have occurred in later years as operational procedures on the lift fell below previous standards

High water levels in the River Weaver on 8 February 1946 resulted in the flooding of the dry well, with the lift being put out of action for a week. The electric motors and cabling to the river gates, and the pump, suffered damage due to immersion. It is not recorded, however, what the cost of repairs was.

In the same year, a contract was awarded to Steeplejack & Contracting Co. Ltd. of Manchester, in the amount of £860, for re-painting the whole lift structure. This was last carried out eight years previously in 1938, which is at odds with Saner's recommendation to the Trustees back in 1910 that the structure should be re-painted '*without fail every third year*'. The painting was commenced in October but, unfortunately, on Sunday 3 November one of the contractor's employees fell from a height of 40 feet, although it is not recorded as to whether the victim survived to tell the tale. The painting works were suspended later in November due to cold weather, which also resulted in the closure of the boat lift and the Trent & Mersey Canal during February and March 1947. The records suggest that, for some unexplained reason, the painting was not recommenced until 9 July.

The disruption of the operation of the lift due to the poor weather led to Marsh giving consideration to the introduction of heating elements to the rubber sealing strips on the gates – the components most affected by the freezing conditions. In September, a quotation for the work was obtained from Fel-Electric Co. Ltd. However, due to cost and the fact that the canal would be similarly affected by cold weather conditions, the Trustees decided to refer the matter to 'the Ministry'. Subsequently, an inspection by Mr. Begg, Chief Engineer of the ICI Alkali Division, on 17 October 1947 resulted in the viability of the proposed use of heating elements being confirmed. No further action, however, seems to have been taken on this matter.

Finally, in June 1949, as a result of the 1947 Transport Act, the Weaver Navigation Trustees were dissolved by the Board of Trade and their functions were taken over by the British Transport Commission. Virtually the whole of Britain's transport system, including the roads, railways and inland waterways, were taken into public ownership. The demise of the Weaver Navigation Trustees was formally announced by the publication of the following notice in the *London Gazette* on 10 June 1949:

'Now therefore the Board of Trade pursuant to Section 24(1) of the Transport Act, 1947, hereby certify that there is no reason for the continued existence of each or any of the said bodies.
J. COWEN
An Assistant Secretary of the Board of Trade.
3rd June, 1949

★ ★ ★ ★ ★ ★ ★ ★ ★ ★

Before closing this chapter, it is worth taking an overview of trading on the boat lift after its conversion to electrical operation in 1908. The annual tonnages passing through the lift initially grew for several years after conversion but traffic began to decline at the onset of the First World War and, apart from a relatively brief resurgence during the 1920's, the downward trend continued up to the start of the Second World War and beyond. The actual recorded tonnages are given in the table below:

RECORDED TONNAGES OF CARGO PASSING THROUGH THE ANDERTON BOAT LIFT 1908 - 1947

Year	Tonnage	Year	Tonnage	Year	Tonnage	Year	Tonnage	Year	Tonnage
1908	163,782	1915	200,606	1922	97,805	1929	140,042	1936	72,497
1909	211,071	1916	166,987	1923	89,756	1930	112,596	1937	75,126
1910	–	1917	133,013	1924	125,922	1931	77,785	1938	69,673
1911	–	1918	105,234	1925	119,962	1932	74,165	1939	66,207
1912	–	1919	72,962	1926	107,134	1933	69,298	1940	45,025
1913	–	1920	90,538	1927	129,960	1934	66,622	1941	64,115
1914	213,771	1921(Dec)	79,802	1928	137,316	1935	74,153	1947	32,624

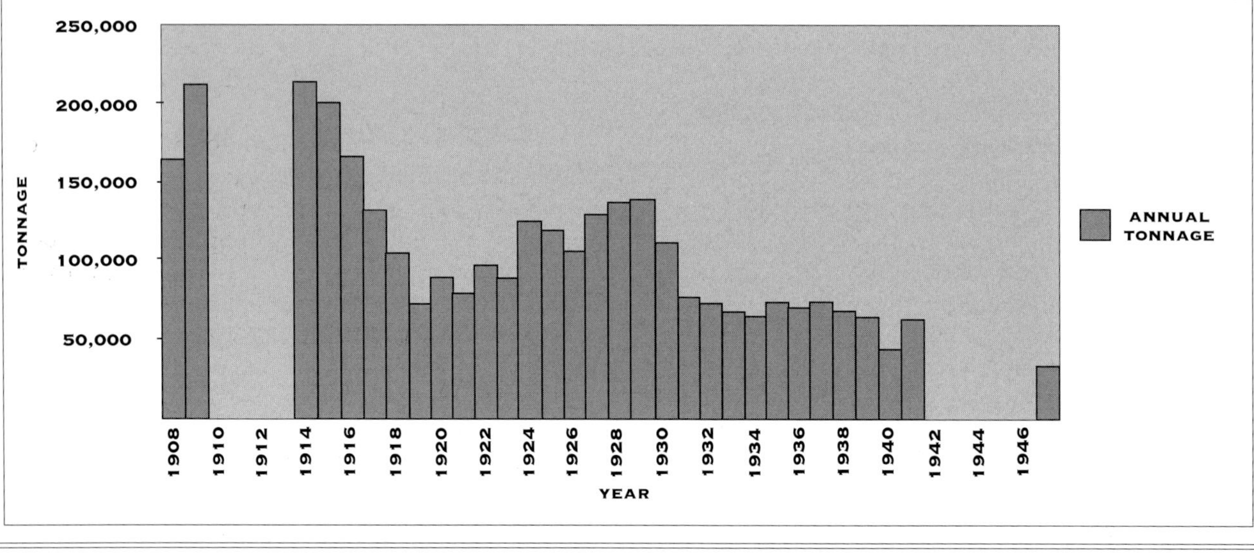

The tonnages for the years from 1910 to 1913 inclusive are not available, but during this period the annual throughput is said to have reached 225,000 tons. Similarly, records are not available for the years from 1941 to 1946, but the tonnage carried in 1947 confirms the ongoing decline in the fortunes of the lift.

The following points are of interest when considering the above figures:

* 1908 - The tonnage for 1908 is relatively low as it was affected by the stoppages for the conversion of the lift to electricity and by the period of single caisson working between May and July 1908.

* 1908 to 1912 - The traffic through the lift continued to grow for several years after its conversion, in spite of competition from the railways.

* 1914 to 1918 - The First World War led to a decline in traffic on the canal system as a whole, which was reflected in the tonnages being passed through the lift. The canals which were owned by railway companies were placed under the jurisdiction of the Railway Executive Committee in order to promote the use of the inland waterways in the national war effort. The employees of these canal companies were given protected status and hence excluded from the draft. The other canal companies and private carriers were not protected in this way and were consequently badly affected, with their employees either enlisting or taking up more lucrative employment in the munitions industry. With an impoverished canal transport system, businessmen, not surprisingly, transferred their goods onto other forms of transport, particularly the railways. The result was that, by the end of the war, traffic passing through the lift had more than halved.

* 1918 to 1929 - Recovery of traffic on the inland waterways after the First World War was hindered by the Depression and the early stages of development of the road transport system in Britain. Competition from roads continued to grow throughout the next few decades. During this period, there were still three salt chutes in use at the Anderton Basin, and one of these remained into the 1940s.

* 1929 to 1947 - A downward trend in traffic volumes was established after 1929, most probably due to the growing dominance of road transport.

* 1940 to 1945 - During the Second World War, the canals once again came under the jurisdiction of the Railway Executive Committee and, as before, the war years witnessed a decline of traffic on inland waterways which was reflected in the tonnages passing through the boat lift. This time many of the employees on the Weaver Navigation were given protected status.

* 1908 to 1947 - The estimated total tonnage of cargo passing through the converted electrical lift, making reasonable assumptions for the missing data, is nominally 4.4 million tons, with an average annual tonnage of approximately 110,000 tons. (It is interesting to note that the average annual tonnage passing through the hydraulic lift during 33 years of operation was approximately 103,000 tons).

* 1949 - At the time of nationalisation, the boat lift was still handling a significant volume of traffic, which probably necessitated both caissons to be operated, together with the employment of two full time staff. It was still, at that stage, though to be operating in profit.

In summary, therefore, the conversion of the lift to electrical operation in 1908 can only be described as a technical and commercial success. Technically it was a very audacious solution to the ongoing problems of corrosion which were being encountered on the hydraulic lift. Commercially, the income over the 40-year life of the converted structure, from both the lift and the Weaver Navigation as a whole, must have more than repaid the initial capital expenditure of £25,000.

INFORMATION SOURCES FOR CHAPTER 8

Steve Astles – *Engineering Supervisor on the Weaver Navigation for British Waterways, North West Region.*

Arthur Dodd – *Labourer, Plant Operator and Dredger Captain for the Weaver Navigation and subsequently British Waterways, Northwich Area from 1938 to 1979. Interviewed on 11 September 1999 at Northwich.*

Brian Haskins – *Area Engineer at Northwich for British Waterways from 1968 to 1988, Engineering Manager for the North West Region from 1988 to 1992 and Chief Civil Engineer at Leeds until 1994. Interviewed at Northwich on 3 September 1999 and subsequent discussions.*

Jack Horton – *Labourer, Joiner and Area Inspector for the Weaver Navigation Trustees and subsequently British Waterways, Northwich Area from 1933 to 1975. Interviewed on 2 September 1999 at Northwich.*

Frank Westwood – *Repair Yard Manager at Northwich for British Waterways from 1964 to 1981. Interviewed at Northwich on 3 September 1999 and subsequent discussions.*

Minutes of the Proceedings of the Weaver Navigation Trustees – 1908 to 1948. *British Waterways' Regional Office, Northwich.*

The Anderton Boat Lift – An Examination of its Impact on Local Trade and Population, and of the Extent to which Trading Figures and Working Conditions Reflect National Economic Trends. *Dissertation by Sharon Rosalind Murray, 1990. Chester College of Higher Education.*

INTERLUDE

A TRIP THROUGH THE BOAT LIFT IN 1959

On Tuesday 8 September 1959, railwayman and photographer Norman Jones, accompanied by his friend and colleague Rodney Hampson (Stationmaster at Mouldsworth Junction), took a trip up the Anderton Lift aboard the British Waterways Maintenance Boat *Hercules*, in the charge of Waterways Maintenance Man John Williams. Norman photographed their progress through the lift and also captured some wonderful images of a working boat family negotiating the lift in the opposite direction. This highly evocative 'moment in time' is presented here over the next few pages as a short interlude to the main story of the boat lift.

The first picture in the sequence, ***right****, is a general view of the approach to the lift from the basin. Note the wooden frame above the eastern caisson gate, indicating it was out of use and single caisson working was in operation. The western caisson has just begun an ascent. Having moored up temporarily to await their turn, the second view,* ***below left****, looks over* Hercules *and shows the caisson descending again, carrying a pair of narrowboats. The caisson's end gate is shown in clear detail As the caisson reaches river level, Waterways Maintenance Man John Williams, having already cast off lines, then prepares to pole* Hercules *out into the basin, ready to take their turn in the lift,* ***below right****. Their 'cargo' consists of Williams' bicycle.*

The river-bound motor and butty pair exit the caisson (at quite a speed, the photographer noted) still 'breasted up' – note the rope lashed across their bows. Note also the distinctively different design of the two vessels; the motor looks to be one of the British Waterways' built 'Admiralty' class boats, with iron hoops over the hold as cloth supports. Their design owed more to a narrow beam lighter but was effective because they could carry around 25 per cent more cargo than a traditional narrow boat, important on the shallow Trent & Mersey Canal. Hercules, *meanwhile, crabs out into the basin, bow pointing towards the caisson. In the background, Seddon's* Weaver Belle *once more features, again looking in very smart trim.*

No time was wasted on these manoeuvres, as this next photograph shows. Hercules *slips neatly into the caisson, nosing her way gently past the emerging pair, whilst the Lift Attendant shares a passing joke with their crew. Note the decorated water cans on the cabin roof of the butty, the neatly stowed sheets along each side of the vessel's hold and the unusually positioned cabin window, complete with curtains.*

Inside the lift looking up, showing the guillotine gate sealing off the aqueduct and the array of wires and pulleys for lifting the caisson. The bow of another boat waiting to descend peeps over the gate, whilst the control cabin can be seen above.

John Williams glances down to make sure the caisson gate is securely seated. In the background appears the giant Winnington Works, by now part of ICI. One of the company's packets – the funnel colours were blue with black tops – is moored to the quay.

Above: *As the caisson rises, more of the quay in the background appears, along with rows of railway wagons, whilst the steam packet can now be seen more clearly. It is either* Anderton *or* Barnton, *these being the only two steam packets operated by ICI at this time. Note the polished paintwork of Williams' craft; the BW men took pride in their vessels.*

Aptly nicknamed 'The Tunnel of Love' by the 1950s boatmen, this is the view looking out from the caisson at the top of its lift, as Hercules *waits for the gates to be raised. Whilst single caisson working was in operation, if there were boats waiting both top and bottom motors and butties had to 'single up' to use the lift, as the aqueduct was not wide enough to allow pairs to be passed.*

Having ascended the lift Norman and his colleague watched the arrival of a British Waterways pair, Dragon *and* Birdswood, *along the Trent & Mersey Canal and followed their progress down the lift.* Dragon *is an Admiralty class motor boat – note the vessel's lower profile and the distinctive hoops showing through the canvas. Notice again the clean and tidy condition of both craft – brasswork polished, cloths, mops and buckets neatly stowed – and the relaxed confident air of the crew as they swing into the top basin.*

A close up of the motor boat Dragon *in the canal basin, with the footbridge from which the previous picture was taken behind. Diesel oil cans stand in front of the cabin and above them is a bar with a heavy steel ring, used for 'roping around' in awkward situations where the rope has to be constantly repositioned, or when 'winding' the boat. We shall see more of the little girl, here with her mother, shortly.*

With the autumn sunlight sparkling on the swirling water, Dragon *and butty* Birdswood *ease their way into the tight fit of the aqueduct. On the left is the sealed off eastern aqueduct with gates down across the entrance. This rear view of the boats, with cabin doors open, shows that, even at this late date and in a nationalised industry, boatmen were still decorating their craft in the traditional manner. More of the sprawling ICI Winnington complex features in the background.*

This magnificent view was described by photographer Norman Jones himself as the most definitive and nostaligic of the Anderton Lift pictures he had taken, capturing as it does, the atmosphere of the late 1950s and the end of the era of narrowboat transport. British Waterways disbanded their fleet in 1963, although unsuccessful attempts were made by others to carry on the industry for a few years after. Dragon *and* Birdswood *wait to enter the caisson, underneath a sign instructing boatmen* WARNING SWITCH OFF ENGINES WHEN ENTERING AND LEAVING THE LIFT. *Above the control cabin, the large board reads* BRITISH WATERWAYS ANDERTON BOAT LIFT.

This superb character study of the little girl, about six years of age, who we saw earlier is actually far more than it first seems. She is steering the boat into the caisson and is not looking at the photographer but concentrating hard on the side pillars behind him, in order to judge the boat's speed. The speed wheel is in the cabin hatchway to her left and she has just given the boat a touch of 'full astern' – observe the creaming wash. Norman Jones described it as an immaculate piece of boatmanship from one so young. Behind her, note the turk's head rope ornament on the tiller headstock of the butty boat. Also the piece of ornamental ropework used to dress her plait and that, despite her responsibilities with the boat, it looks like she's been at the chocolate!

A final glimpse of the extended boat family – boatman, engineer, their wives, two children and the dog – with their pair of boats is provided in this dizzying view down the lift. With little chance to chat, the photographer regretted he was not able to get any more details about the family except the fact they were bound for Ellesmere Port.

All photographs courtesy Norman Jones FSCA

The narrowboat Ivy *moored outside of an unidentified sister craft at the Mersey Weaver & Ship Canal Carrying Company's wharf in the 1950s, with the boat lift looming through the haze in the background, in a picture which epitomizes Tom Rolt's quote on the opposite page.*
Courtesy The Boat Museum, Ellesmere Port

CHAPTER 9
OPERATION OF THE ELECTRICAL LIFT (PART 2)

'Towering above the shipping on the Weaver, the great black bulk of the Anderton lift makes an impressive sight, but never more so than in early morning or late evening when it looms indistinctly through the mists of the river and the fumes of the factories.'

Extract from The Inland Waterways of England *by L.T.C. Rolt, 1950*

In 1947, the Labour Government brought out the Transport Act which, within a year or so, led to the nationalisation of Britain's transport system, including the roads and railways, together with the canals and river navigations. The newly nationalised organisations came under the authority of the British Transport Commission, with day-to-day operational control being vested in various executives. In consequence, in June 1949, the inland waterways system, including the Weaver Navigation with the Anderton Boat Lift, was put under the control of the Docks & Inland Waterways Executive.

The Act brought to a close over two hundred years of operation of the Weaver Navigation since 1721. Apart from the early years under the original Commissioners, the Weaver Navigation was hugely successful and this in turn, due to the special clauses in the 1721 Act, brought major benefits to the local communities, with surplus revenue being transferred to the County of Chester for the improvement of roads and bridges, and other worthy causes. There is no doubt that the eighty-six gentlemen who obtained the first Weaver Navigation Act had achieved their objective of opening the River Weaver to navigation on condition that '*the profits were used for the common good*'.

The nationalisation of the transport system by Clement Attlee's Labour Government could easily be mistaken as a benevolent attempt to rescue the country's rail and canal systems, which were suffering under increasing competition from road transport. Of course, things are never that simple, and it is probably more accurate to describe nationalisation of the waterways as a politically motivated act.

It should be remembered that the years between the two wars had seen a steady decline in the fortunes of the inland waterways of Britain. It was a similar situation for the Weaver Navigation and, hence, the Anderton Boat Lift, which continued to lose commercial traffic to the ever-expanding road network. At the time of nationalisation, traffic on the Weaver Navigation was at an all time low, which is amply demonstrated by the recorded annual tonnage through the boat lift in 1947 of 32,624 tons. In comparison, the peak reached a few years preceding the First World War was close to 225,000 tons. Despite this, the Weaver Navigation as a whole was in all probability still operating in profit in 1949, which was perhaps one of the reasons why the inclusion of it in the nationalisation process was fiercely, but unsuccessfully, contested by the Trustees.

The main target of nationalisation was undoubtedly the road and rail systems. However, due to large parts of the canal system having been bought by railway companies during the boom years in the second half of the 19th century, much of the inland waterway system was brought into public ownership at the same time. For whatever reason, the Weaver Navigation and presumably other river navigations suffered the same fate. A tongue-in-cheek tale on this subject, prevalent around the years after 1949, was told by Brian Haskins (Area Engineer at Northwich for British Waterways between 1968 and 1988). The story went that the Transport Minister at the time of nationalisation, on being asked by his civil servant what he wanted done with the canals, responded by saying, '*What! Have we got them as well?*', seemingly not realising that the waterways had been acquired as well.

Regrettably, and perhaps not surprisingly, the process of nationalisation did not cure the financial difficulties of the waterways – in fact it probably did much to exacerbate them. Whilst various re-organisations of the management structures took place, there was seemingly no new money available, money which, by that time, was desperately needed to address years of under-expenditure on the canals. On the contrary, it would seem that pressure was applied on the newly formed public bodies to reduce operational costs, to address the continuing loss of income as more and more traffic was transferred from canals to roads. The thirty or so years following nationalisation, leading up to the eventual closure of the Anderton Boat Lift in 1983, were to be difficult times for the inland waterways in general, and the Weaver Navigation in particular.

Whilst re-structuring of the inland waterways undoubtedly caused turmoil in the upper echelons of management of the new public bodies, it is reasonable to assume that life went on more or less as usual in the lower ranks. Operational staff working on the Weaver Navigation

Certificate

OF THE BOARD OF TRADE

Pursuant to Section 24 (1) of the Transport Act, 1947 the Board of Trade have issued the following Certificate.

Whereas the Board of Trade are satisfied in relation to each of the bodies specified in the schedule hereto:—

(1) That the body have, in accordance with Sections 12 to 23 inclusive of the Transport Act, 1947, collected and distributed any moneys receivable by them in respect of any agreement disclaimed by the British Transport Commission (including the appropriate proportion of any moneys received by any joint committee or joint body from any person in respect of any agreement of that joint committee or joint body which is disclaimed by the said commission);

(2) that the body have distributed any moneys paid to them by the said Commission;

(3) that the body have complied with any directions given to them by the said Commission for the purpose of securing that the ownership of any property or any right is effectively transferred to the Commission; and

(4) that the said Commission do not desire to give any further such directions.

Now therefore the Board of Trade pursuant to Section 24 (1) of the Transport Act, 1947, hereby certify that there is no reason for the continued existence of each or any of the said bodies.

J. COWEN,

An Assistant Secretary of The Board of Trade.

3rd June, 1949.

The above Certificate was published in the LONDON GAZETTE on 10th June, 1949, from which date the Weaver Navigation Trustees were dissolved.

SCHEDULE

The Mersey Railway Company,
The North Devon and Cornwall Junction Light Railway Company,
The Southport and Cheshire Lines Extension Railway Company,
The Great Central and Midland Joint Committee (Lessors).
The Weymouth and Portland Railway Company,
The Great Western and Great Central Railways Joint Committee, (Lessors),
The East Kent Light Railways Company,
The Company of Proprietors of the Coventry Canal Navigation,
The Leeds and Liverpool Canal Company,
The Staffordshire and Worcestershire Canal Company,
The Aire and Calder and River Dun Navigations Joint Committee,
The Whitechapel and Bow Railway Company,
The Southern Railway Company,
The Weaver Navigation Trustees,
The Sheffield and South Yorkshire Navigation Company,
The Company of Proprietors of the Birmingham Canal Navigations,
The Company of Proprietors of the Herefordshire and Gloucestershire Canal Navigation,
The Eastern and Church Hope Railway Company,
The Undertakers of the Aire and Calder Navigation.

*The Certificate from the Board of Trade, **left**, advising of the dissolution of the Weaver Navigation Trust, and other private canal and river navigations, as published in the* London Gazette *on 10 June 1949. The Schedule, **above**, lists the concerns affected.*

Above and left, opposite: *Two more views taken in summer 1948 at the time of Sir Cyril Hurcomb's visit to the lift. The photograph left, shows the party standing alongside the structure, watching the Fellows, Morton & Clayton narrowboat, which they had previously seen entering (see page 8), ascending in the western caisson. Behind them, the smartly turned out steam packet* Weaver Belle, *unusually without her dumb barge* Gowanburn, *awaits unloading. Operated by salt producers Henry Seddon & Sons, she was built in 1900, and had steel frames and timber planking. The photograph above, taken a short while later, shows the inspection party back on the footbridge again, extreme right, whilst the* Weaver Belle *transfers her cargo, by means of her derrick, to a waiting Mersey Weaver narrowboat (the cargo may be bagged china clay, bound, via the lift, for the Potteries, or bagged soda ash, produced locally by ICI and used in the manufacture of various products, such as soap, leather, paper, glass and bleach, or alternatively bagged salt). This side-on view of the lift reveals a far more graceful structure than in the many front and three-quarter views.*

Both Neil Parkhouse collection

would have been transferred to the nationalised authority, initially at least, without significant change in their duties. Similarly, day-to-day life at the boat lift was probably unaffected by nationalisation.

★ ★ ★ ★ ★ ★ ★ ★ ★ ★

It should not be forgotten that, by 1949, the elements of the original hydraulic boat lift which had been retained after conversion to electricity were over 70 years old, whilst most of the remainder was over 40 years old. The passing years and the badly polluted environment at Anderton had both made their indelible marks on the structure. The worsening condition of the boat lift seems to be confirmed by Arthur Dodd who, when acting as temporary assistant on the lift in the early 1950s because George Beech, the Lift Attendant, had injured his ankle, stated that:

'I've been up top with him [Beech) *in the* [control] *building, and I can remember when he was lifting the gates up to let the craft in* [to the caissons] *and how the floorboards would shake. It* [the lift] *was getting a bit dodgy by then.'*

Curiously, however, the components worst affected by the ravages of corrosion seemed to be those installed in 1908, whilst the older 1875 components had generally fared better. This phenomenon is largely explained by the state of materials technology at the turn of the century. The principal structural elements in the 1875 lift were the vertical columns and the hydraulic rams beneath the caissons, all of which were designed to act in compression. These elements were fabricated from cast iron, as this material was known to perform well in compression and could, by the end of the 19th century, be produced relatively cheaply and to a consistently high quality. Another advantage of cast iron was its high resistance to corrosion. By 1908, however, the construction industry was in the throes of adopting steel, a relatively new material which, unlike cast iron, was strong in both compression and tension. Steel therefore offered the designer greater freedom of structural form, permitting the use of horizontal beams for large spans where arch construction would previously have had to be adopted. Steel, therefore, was seemingly the ideal material for the 1908 conversion of the boat lift and was used for the beams in the machinery platform and also for the supporting 'A' frames. The main drawbacks to the use of steel at that time were its vulnerability to corrosion (compared to cast iron) and production methods, which were still being developed and improved. As stated by Wallace Evans & Partners in their report entitled *Anderton Barge Lift Rehabilitation - Presentation of Alternative Schemes*, prepared in 1984 after closure of the boat lift:

'No such quality controls [for steel production when

Although a well known picture, this view of the Mersey Weaver pair Walton *(motor boat) and* Darlaston *(butty) exiting the eastern caisson at the top of the lift certainly merits inclusion here. The year is 1956 and the boats are in the charge of Mr. and Mrs. Jim Evans, who are heading south with material for the Potteries – note also the bicycle stored amidships! It will be noted, too, that* Darlaston *is still in Anderton Company colours, although they had sold out to the Mersey Weaver concern two years previously. The latter were themselves taken over two years later by British Transport Waterways.* Mrs. Joy Parkinson collection

compared to that of cast iron] *were in operation at the end of the 19th century and the construction boom of the decades straddling the turn of the century resulted in hurried and often poorly controlled manufacture of steel products with a range of properties.'*

This is still evident on the structure today by comparing the condition of the cast iron guide columns of the 1875 hydraulic lift with the steel columns in the 'A' frames of the 1908 conversion. The latter, now over 90 years old, are very badly pitted, particularly at the northern end of the lift where the columns have been exposed to the leakage of canal water from above. The 1875 columns, although exposed to the same conditions and for some 30 years more, are in remarkably good condition for their age. Similarly, the steel framework of the 1908 machinery platform appears to have suffered more from the ravages of time than the wrought ironwork in the 1875 aqueduct and caissons.

★★★★★★★★★★

Apart from the adverse financial consequences, the Weaver Navigation suffered in other ways from nationalisation. Prior to 1948 the Navigation had been a self-contained section of the nation's transport system, run on a private basis by the Weaver Trustees, with each year's excess profits going to the county. In the early years the Navigation, by necessity, employed local men and there would have been many cases of successive generations from the same family working on the Weaver. This was very much the situation for the Beech, Carter, Platt and Shaw families. In Arthur Dodd's case, his uncle was a Foreman and lock gate worker, his half-brother, Jack Horton, ended up as a Senior Area Inspector and his cousin was Captain of a dredging hopper called *Dolphin*. Arthur himself ended his working career in 1985 as Captain of the bucket dredger *Vera Barton*.

Over the 200 years or so of its existence, the Weaver Trustees gained a reputation for being caring and reliable employers who had strong links with the local community. The men responded accordingly with loyalty and commitment. In Arthur Dodd's words:

'When I worked on the Weaver Navigation, everybody got pride in their work, whatever they did. I think we lost a lot of our pride [after nationalisation]. *It was part of our lives and part of my life and my family's life. It was the same with the other families. We took an interest. I put many hours in for nothing.'*

Nationalisation resulted in the assimilation of the Weaver Navigation into the rest of the waterway network and this gradually led to the loosening of the bonds between employers and employees. Some men who, the year before, had worked solely on the River Weaver were given duties on the nearby Trent & Mersey Canal and vice versa. Similarly for managers, who were once based at Northwich but after nationalisation were given responsibilities over a greater area of the waterways system, or even re-deployed to other parts of the country. Travelling long distances to work was now not only possible but had become for many a necessity. In time, these changes led to the breakdown of the special relationship that existed amongst the staff of the Weaver Navigation, with unfortunate consequences.

★★★★★★★★★★

Soon after nationalisation, probably in 1953, George Beech, one of 'old school' employees on the Weaver Navigation, was transferred from his work as plant operator to the post of Lift Attendant at Anderton, a position which he held until 1967. In the early years of his employment on the lift, it is likely that Beech had an assistant who worked the gates at ground level. However, due to the declining volume of traffic and the need to reduce costs, it would not have been long before he was required to work the lift single-handed, without a regular assistant. Working on his own, Beech would have had to carry out all the tasks necessary to operate the lift. In transferring a boat from the canal to river level, he would first have assisted the boat into the caisson from the aqueduct and then climbed the stairs to the control cabin to lower both sets of gates (aqueduct and caisson). Next he would leave the cabin and operate the valve to drain the chamber between the gates, then close it and return to the cabin to instigate the descent of the caisson, making sure beforehand that the safety pawl had been applied. While the vessel was descending, Beech would go down the stairs to ground level where, once the caisson had reached the end of its travel, he would fill the chamber between it and the river gate, finally raising the gates allowing the boat to move out into the river basin (the extent of the movement of the caissons was presumably controlled by travel-limit switches). Consequently, when traffic was heavy Beech was

George Beech, the Lift Attendant in the 1950s and 60s, in full uniform whilst on the machinery platform of the boat lift, circa 1960. Courtesy British Waterways

kept fully occupied. There would, of course, be someone who would stand in for him on his days off.

Jack Horton started working for the Trust at around the same time as Beech in the early 1930s and they ended up being friends and neighbours, living as they did in adjacent Trust cottages at Hunts Lock, in Northwich. Jack started work on the Weaver Navigation at the age of 25, as a labourer and, in the early years at least, his responsibilities included the repair and replacement of the timber components on the lift structure. According to Jack, George Beech was a traditional type of employee who took pride in his work and was fiercely independent. He maintained a clear distinction between his duties and those of other workmen, and was said to be very selective as to whom he would allow into the control building. Beech was respected by his colleagues and managers alike. In Jack's words:

'They knew the man, and they knew they could trust G.T. Beech, with their life never mind with the Lift. There was nothing go wrong there without he wasn't on the phone [to the Northwich Repair Yard] *and that's it and they'd get out to him straightaway. But it was surprising, they never seemed to have any trouble, with George.'*

This view of George was very much confirmed by Arthur Dodd:

'As I say, George Beech was an excellent man. He knew all about electrics and everything [on the lift]. *He was a wonderful man. And he wasn't a tradesman, he started as a boy. His father was a captain on one of the wood hoppers and his brother was also on a dredger. Like I say, a family concern, but George was the brightest of the lot of them. He was a clever chap, George. He was a very good friend of mine.'*

Frank Westwood, who was the Repair Yard Manager at Northwich between 1964 and 1981, had similar opinions about Beech:

'He [George] *was such an excellent man that if I sent someone down* [from the Repair Yard to work on the lift] *George would be guiding them because of his knowledge of the lift.'*

The presence of Beech, therefore, as Lift Attendant at Anderton over the fourteen or so years after nationalisation, must have done much to slow the inevitable process of deterioration of the boat lift, caused by lack of funds and corrosion.

★★★★★★★★★★

Another change occurred in 1953, when the inland waterways came under the jurisdiction of the newly formed British Transport Waterways but still in the overall control of the British Transport Commission. By this time, traffic through the lift, according to Arthur Dodd, had dwindled to just four or five commercial boats a day, with little or no leisure craft. The 1950s and 60s marked the quietest period in the life of the boat lift, when commercial carrying was coming to an end but the value of the canals for leisure purposes was yet to be realised.

In 1958 major repair works were carried out to the lift, which included replacement of the aqueduct gates, caisson gates and the river gates, all of which were severely affected by the ever present problem of corrosion. During these repairs, several badly corroded wrought iron components were replaced by steel. The overall cost of the works was nominally £8,000.

Around 1962 the locks at Runcorn, which linked the Trent & Mersey Canal with the River Weaver via the Bridgewater Canal, were finally closed to traffic, with the result that from then on the Anderton Boat Lift was the only means of communication between the two waterways. Presumably this resulted in a temporary increase of traffic passing through the lift.

In 1962, the Aqueduct was re-painted by British Waterways' staff and the lift structure was re-painted by outside contractors.

Yet another re-structuring of the inland waterways took place as a result of the 1962 Transport Act, which led to the abolition of the British Transport Commission on 1 January 1963 and the creation of the British Waterways Board. The Weaver Navigation, and hence the Anderton Boat Lift, has been under the control of the British Waterways Board, now referred to simply as British Waterways, ever since.

Commercial traffic on the Weaver Navigation and the Anderton Boat Lift continued to decline. The boat lift was being operated by George Beech generally on his own but by the early 1960s it had become normal practice to operate the lift under continuous single-caisson working. The two adjustable frames at the river basin entrance, onto which the caissons sealed when at river level, were fixed at different heights so that one of the caissons was operated when the river was at or slightly above its normal retention level, whilst the other caisson would be used with higher river levels. This negated the need to regularly adjust the position of the sealing frames according to the state of the river, which would otherwise have been necessary to avoid under or over-filling the caissons.

Economies were still being made on the Weaver Navigation, with reductions in staff both at the Northwich Yard and on the river. By 1964, British Waterways' fleet of narrowboats based at Hayhurst Yard in Northwich was in the process of being sold off, presumably as they were no longer considered to be profitable. The sale of the fleet would undoubtedly have caused a significant reduction of traffic passing through the lift. Also, working practices began to change, due to lack of traffic, with reductions in manning on structures and those staff remaining being required for duties elsewhere on the Navigation. This had the regrettable consequence of reducing the men's affinity with individual structures, taking them away at times when previously they would have been available to carry out routine maintenance work.

Whilst the Area Engineer at Northwich had overall responsibility for the operation of the Weaver Navigation including the Anderton Boat Lift, he relied on fitters from

Above: *The 1950s were the twilight years of waterborne transport, particularly on the narrow canals, but some of the traditional carrying areas could still provide scenes of apparent busyness. British Waterways Hayhurst Repair Yard is seen here in the late 1950s, with a selection of working narrowboats on view. Some of these are probably awaiting work, however, rather than in the yard for repairs. Note the electric lights on the cratch of each boat, for tunnel use rather than night treavel.*

Right: *The narrowboats* Hereford *and* Parrot *moored in front of the offices of British Waterways Southern Carrying Fleet near the lift, which can just be seen beyond the footbridge. The basin at the top of the lift also appears to hold four or five vessels in this circa 1960 photograph.*

Right: *Another 'crowd' scene, on the Trent & Mersey, but again these boats are probably awaiting or on route for orders. Although this may well be a Sunday view, it would still take a day or two after this to find work for all these craft. There was still plenty of work but it was dwindling and time spent moored up was on the increase.* Dory *and* Alder *are the only two identifiable craft but note the appearance of all the boats; decorative paintwork, polished brass, the ornamental ropework on the tiller headstock of* Dory *and the overall impression of cleanliness.*

All courtesy The Boat Museum, Ellesmere Port

Above: *A young lad leans on the cabin of an unidentified motor boat in another late 1950s photograph. It is lashed to the butty* Turtle *and they are moored in the canal basin at the top of the lift, awaiting a load. It would appear that, as the aqueduct gate is lowered, the western caisson is not in operation.* Courtesy The National Waterways Museum, Gloucester

The eastern, redundant, arm of the river basin photographed on a typical November day in 1956, shortly before it was back-filled. It had been used as a dumping ground for old Mersey and Weaver flats, which had latterly been used as dumb barges on lightering duties. A packet can be seen moored to the quay in the background. Courtesy Edward Paget-Tomlinson

There was still work to be had and cargoes to be transshipped, as this July 1956 photograph of Anderton Basin shows. Weaver Belle *and her dumb barge* Gowanburn *box in a narrowboat whilst transferring cargo, which again is either bagged china clay, soda ash or salt. Another narrowboat appears to have completed loading ahead of the packet's bow. Although not used for some time by the date of this view and in fact soon to be demolished, one of the salt chutes survives on the right. Note* Weaver Belle*'s coggy boat astern.*
Courtesy Edward Paget-Tomlinson

A cargo of what looks to be paving slabs seen being transshipped at Anderton into the motor boat Cormorant *and her unidentified butty, again in the mid 1950s. The boat family watch proceedings with quiet interest. Note the bicycle, a useful boatman's aid; whilst providing transport to any nearby town for shops or the like, its main use was for 'lock wheeling', whereby a crew member would get ahead of the boat to the next lock and prepare it for a quick passage, providing there wasn't a queue of course.* Courtesy The Boat Museum, Ellesmere Port

the Northwich Yard to carry out repair works with respect to the mechanical and electrical equipment on the lift. This included maintenance and replacement of the lifting ropes which, because of inadequate access along the machinery deck, necessitated the use of bos'un's chairs to suspend men over the side of the lift. This was particularly necessary to gain access to the safety rope anchorages and was an operation fraught with danger; it was consequently disliked by many of the men.

Also, at that stage, British Waterways still retained a team of men to paint the various structures on the local waterways, including the lift, a task that again required the use of bos'un's chairs.

Around this time, presumably due to re-organisation within British Waterways, the Regional Engineer at Leeds began to be involved with the Anderton Boat Lift, being called in, as and when necessary, by the Area Engineer to give specialist structural assistance. Consequently, in 1965, after a statutory inspection of the lift, the Regional Engineer commissioned consulting engineers, Husband & Co, of Sheffield to assist with remedial works. The statutory inspection identified the following faults:

'1. Caisson and gate hoist motors and brake units are in a rusted condition.

2. Sections of two main joists carrying overhead gear and structure are in an advanced state of corrosion

3. One caisson gate joint at the aqueduct end is wasted in places.'

British Waterways' fitters from the Northwich Yard dealt with the first of these problems and, in June 1965, Husband & Co. were called in to address the last two. At the same time, they were instructed to carry out a condition survey of the remainder of the lift structure and advise on levels of maintenance. The consultants' report in August 1965 gave twelve recommendations, which included repair and replacement of badly corroded sections of steelwork (particularly with respect to the outer main girders of the machinery platform); re-painting of the whole structure last carried out in 1962; replacement of defective timber planks in the various walkways; overhaul of the gate inter-space valves with motorisation of the river gate valves; and the installation of a centralised lubrication system. The latter two recommendations appear to relate to a proposal to

A rather forlorn looking boat lift, photographed in January 1969, with the low winter afternoon sun lighting its features quite beautifully. The paintwork looks in good condition but it belies the corrosion problems the structure was suffering with by this date.
Courtesy The National Waterways Museum, Gloucester

enable the river gate valves and the lubrication system to be operated from the control building, thereby reducing the amount of clambering up and down the Lift Attendant was required to do. Staff from the Northwich Yard probably carried out most of the works identified but it is possible outside contractors were employed to carry out the more specialised tasks.

It is interesting to note that the three inner main girders of the machinery platform were, when compared to the outer girders, found to be in relatively good condition. The consultants suggested this was partly due to the amount of grease which fell onto the inner girders from the machinery above and also they were less exposed to the weather. In addition, Husbands' report recorded:

'The traffic using the lift consists mainly of barges carrying material from the Potteries, together with pleasure craft. During a recent week 32 vessels used the lift. We were informed by the lift attendant that this is an average figure that would be increased in the holiday season by pleasure craft. The lift is not normally operated during the hours of darkness or at weekends after noon on Saturday.'

★ ★ ★ ★ ★ ★ ★ ★ ★ ★

By 1967 traffic through the boat lift had dwindled still further. It was consequently decided that George Beech, who had been the Lift Attendant for fourteen years, would be more usefully employed as a plant operator, the position he had occupied prior to 1953. Regrettably, due to the continuing pressure to reduce operational costs, the replacements on the lift tended to be less experienced and they did not necessarily take as much care with the lift as Beech had. Another aspect which should not be overlooked was the growing influence of trade unions within British Waterways and particularly on the River Weaver. The presence of the unions at that time, whilst undoubtedly doing much to protect the conditions of employment of their members, sometimes created bad feeling between management and staff, which in turn resulted in some of the men being less committed than they might otherwise have been. The structures on the Weaver Navigation, including the Anderton Boat Lift, may have suffered as a result.

Five years later, the condition of the boat lift was again giving concern to the Area Engineer. As before, Husband & Co. were brought in under the direction of the Regional Engineer to carry out a detailed assessment of the structure. In the introduction of the report presenting their findings, dated February 1972, Husband & Co. rather tactfully stated:

'We understand that the lift was inspected in 1958 and that certain members of the structure were replaced or repaired at that time. We carried out an inspection in June 1965 and a number of recommendations for repairs were made in our report of 6th August, 1965. Some of these recommendations have not been implemented.'

Whilst it is not recorded which of their earlier

A British Waterways Maintenance Boat entering the western caisson via the aqueduct from the Trent & Mersey Canal in January 1969. Again the paintwork of the lift looks good. Courtesy The National Waterways Museum, Gloucester

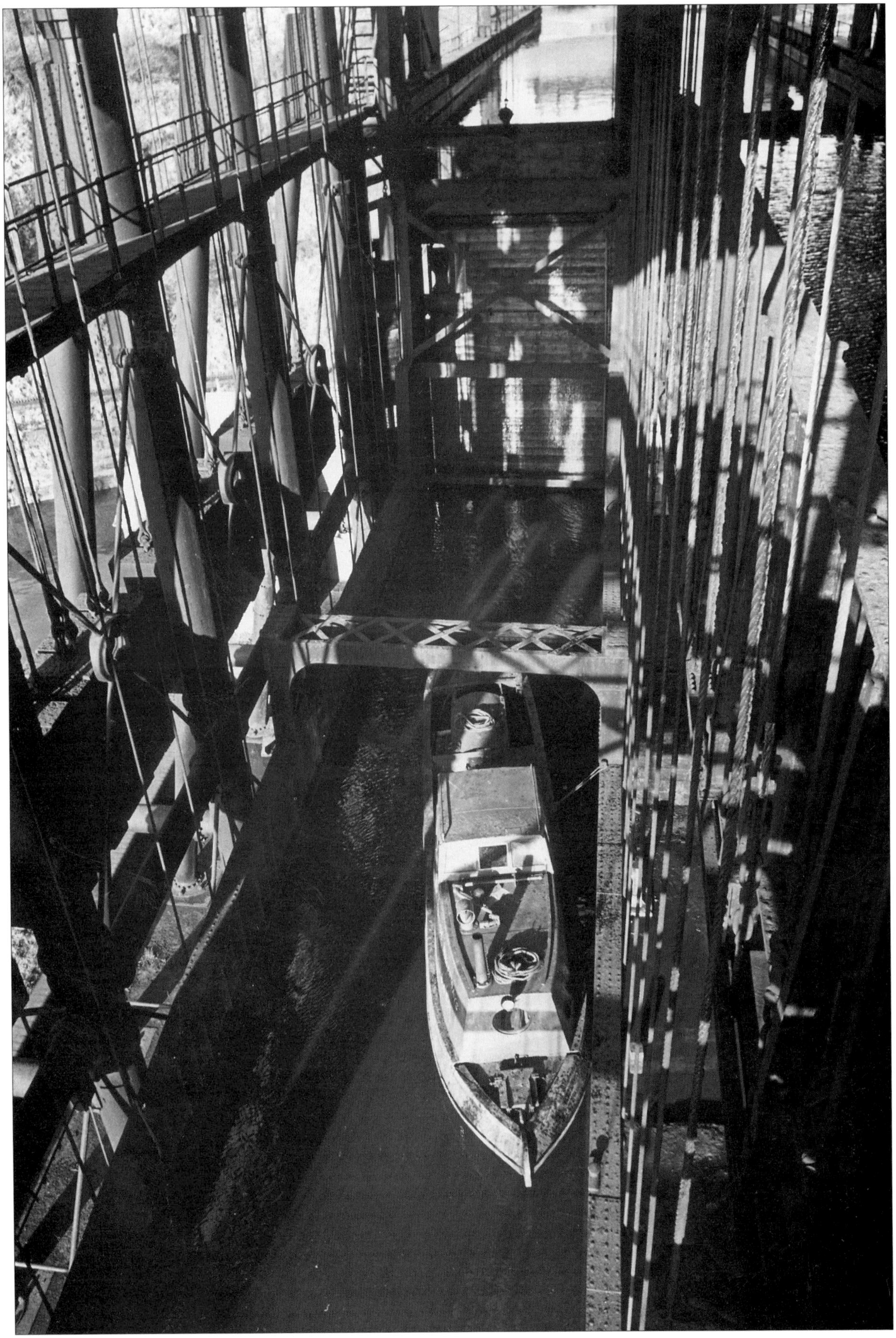

The maintenance boat descending to river level. This and the previous two views were taken by photographer Derek Pratt for British Waterways, possibly because some work had recently been done on the lift. The low sunlight has lit the interior of the structure to good effect. Courtesy The National Waterways Museum, Gloucester

recommendations had not been implemented, it seems likely that the motorisation of the river gate valves and the provision of a centralised lubrication system were overlooked.

The report goes on to advise that the boat lift was in very poor condition and numerous recommendations were made for urgent repair works. The consultants' report suggests that the deterioration had been a gradual process over the seven years since their previous inspection, which would almost certainly have been principally due to the age of the structure but, by this time, lack of adequate maintenance and repair on the aging structure could well have exacerbated the situation. The full extent of the problems associated with the lift was identified in the consultants' report by the following statement:

'The structure is badly corroded at many places and unless steps are taken in the near future to arrest this corrosion and to repair the damaged sections then we are afraid that there may be a major collapse.'

The report identified twenty-two separate areas in which work was considered to be necessary in order to return the lift to a safe working condition; a noticeable worsening of the situation since the previous inspection seven years earlier. The first, and perhaps most significant, of the recommendations was for detailed inspection of and extensive repairs to the aqueduct and caissons, which were particularly badly affected by corrosion. Other suggestions included the removal of the lifting gates to the caissons and aqueducts followed by their cleaning, repair and re-painting; extensive repair and replacement of various timber elements, including fenders, walkways, baulks and facings; cleaning, repair and replacement of various items of the mechanical and electrical operating mechanism; regular greasing of gear wheels in winter and summer, and repairs to the masonry retaining walls of the dry well. Of particular interest was the statement that:

'To drain the space [between the aqueduct and caisson gates] *there is a valve that discharges the water through a pipe into the dry well below. The valves have become so rusted and worn they only work occasionally. The method used at present to drain the space between the aqueduct and caisson is to drop the caisson slightly so that the seal between the aqueduct and the caisson is broken and water is allowed to spill out over the supporting structure below.'*

This is the first recorded instance of, presumably, a relatively recent practice which continued into the later years of the life of the boat lift and that would eventually have almost disastrous consequences. No doubt it began as a way of overcoming difficulties with the drainage valves but over the years, due to the saving of effort involved, the practice gradually became the norm. Jack Horton's description of this was '*The lazy man's way*' or '*smart action*'.

Other points of interest in Husbands' report were:

* The outer girders of the machinery platform were now in such poor condition that they might need to be replaced.

* The gate guides on the aqueduct and at the northern end of both caissons were found to be in poor condition being badly worn and corroded.

* Serious rusting was identified on components that had been installed as recently as 1958. The consultants' explanation of this was the poor quality of the paint used at the time, but could also have been a function of poor preparation prior to painting and perhaps lack of adequate maintenance. It would seem that the practice of grit-blasting to thoroughly prepare metalwork before painting had then only recently been adopted on the Weaver Navigation. Prior to this, the surfaces had presumably been scraped and wire-brushed only; a more painstaking and generally less effective procedure.

* The Lift Attendant advised that '*if a coaster collides with the western bank of the dock, the whole of the lift structure vibrates.*' The consultants recommended that investigations should be carried out, and advised that '*some strengthening of the main wall might be required.*'

* In January 1972, only two barges used the lift.

Husbands concluded their report by stating that:

'. . . subject to the possibility of other defects being discovered during repairs, the items in this report represent the probable minimum amount of work necessary to keep the lift in service, or even to prevent it from becoming a public danger if out of service.'

It is likely that most, if not all, of the recommendations were acted upon by British Waterways. Once again, many of the tasks would have been carried out by fitters from the Northwich Yard but very likely with outside contractors being employed for the larger and more specialised tasks, including the replacement of the outer girders of the machinery platform. However, some of the incidents which occurred in subsequent years cast doubt on the effectiveness of some of the remedial works.

* * * * * * * * * *

To reduce running costs still further, the in-house painting team on the Navigation was gradually run down, with the result that, by the early 1970s, much of the painting of structures such as the Anderton Boat Lift, was being carried out by contractors under the supervision of the Regional Engineer. It was possibly around this time that tolls were re-introduced for the use of the lift, presumably to take account of the growth in leisure craft and the decline in commercial traffic. It is not known what level the tolls were originally set at but, by 1981, the cost of a one-way trip for a boat was £7.50p.

* * * * * * * * * *

In 1976, the Anderton Boat Lift was scheduled as an 'Ancient Monument' by English Heritage under the Town & Country Planning Act. The schedule, rather prosaically, describes the monument in the following way:

'A boat lift connecting the River Weaver to the Trent and Mersey Canal. Built in 1875 as a hydraulic lift, it was converted to electric power during the 20th century. A wrought iron aqueduct carries the canal, which is 50ft above the river, to the two independently operating tanks in which craft are raised and lowered. The structure is still in use and in fair condition.'

★★★★★★★★★★

Towards the end of the 1970s, it is possible the competency of the Lift Attendants had led to a deterioration in the standard of routine maintenance of the lift and the adoption of further poor operating procedures. As previously mentioned, the draining of the chambers between the aqueduct and caisson gates by lowering the caisson by a foot or two became commonplace rather than occasional. Also around this time, the Lift Attendants had adopted the practice of freeing the catch systems, which enabled the aqueduct and caisson gates to be raised and lowered together, by hitting them with a block of wood. The condition of the catch systems had not specifically been remarked upon in the 1972 inspection report but by then they were obviously in need of maintenance and repair.

The poor operational and maintenance practices which had developed since the retirement of George Beech is highlighted in a memorandum dated 30 April 1981, written by Brian Haskins. He advises of a repeat of an incident that previously occurred in 1944. The latest incident, which happened around 1978 and could have had very serious consequences, was reported as follows:

'It appears that . . . . the downstream Weaver end gates leaked badly and either, unknown to the Lift Attendant or if it was known to him, he took the risk. The gates continued to leak fairly badly and of course the tank (caisson) became lighter and lighter in weight as it ascended. This caused the counterbalance weights to overtake it and it began to accelerate rapidly as it ascended. This acceleration apparently did not cause the ropes to slip on this occasion, but the friction between the ropes and the pulleys caused the main pulleys to accelerate thus driving the worm and the worm wheel and in turn the electric motors. This caused the motor to accelerate and eventually I understand the motor burnt out, but the Lift also came to the top very quickly indeed causing all the machinery to whine extremely loudly and this fact was heard in the Anderton Depot by the Depot Superintendant and his Foreman and men and they apparently commented on it at the time.

The above incident apparently caused no particular structural or mechanical damage except that the motor burnt out and it was replaced in due course. It was not thought fit to report this incident and it has not come to my attention on a previous occasion nor was it reported to this office to my knowledge.'

As an aside, it is rather strange that the original design of the boat lift by Saner had foreseen the need for safety pawls to protect against damage to the motors while the caissons were being lowered but did not include similar safeguards for when the caissons were being raised.

The type of incident described by Haskins could be one reason for the relatively frequent motor failures which occurred around this time. Another explanation, as put forward by Frank Westwood, was the length of time that the motors were by then standing idle, getting cold and damp, due to the low volume of traffic.

It is thought that, in 1978 or 1979, further contract painting of the lift was carried out under the supervision of Husbands. According to Brian Haskins, however, this was not very successful, as the preparation work only consisted of scraping and wire brushing, and not shot-blasting as is the norm today. It is probable that, during the course of these works, the safety ropes were removed to enable the winding drums to be painted. It was, perhaps, a failure to adequately tighten the rope fastenings on completion of these painting works that contributed to a very serious incident a few years later.

★★★★★★★★★★

Although there appears to be no continuous record of volumes of traffic through the lift since nationalisation, occasional glimpses are available which clearly indicate the decline in commercial traffic and its gradual replacement by pleasure craft. By the mid 1970s pleasure traffic had totally replaced commercial traffic. Monthly records for a three year period from 1977 indicate how, with pleasure traffic only, boat movements on the waterways had become seasonal in nature:

NUMBER OF CRAFT USING THE ANDERTON BOAT LIFT

	1977	1978	1979
January	0	0	1
February	0	0	3
March	12	0	3
April	49	10	22
May	47	24	72
June	87	62	81
July	175	94	143
August	134	92	109
September	47	32	148
October	6	8	73
November	3	0	10
December	0	0	0
Total	560	322	665

One reason for the relatively large number of lifts carried out in 1979 was the decision by the Inland Waterways Association to hold their national rally at Northwich that year.

By 1979, it is probable that the boat lift had been operating at a financial loss for many years, possibly since the mid 1960s, with income from boat tolls being significantly less than the cost of operating and maintaining the ageing lift structure. It is on record that the income from tolls in 1979 was just £3,290, whilst the maintenance bill was nominally £21,000.

★★★★★★★★★★

Perhaps not surprisingly, given the deteriorating condition of the lift structure and the poor working practices that had been adopted, several potentially serious incidents occurred around this time. Two separate incidents resulted in caissons, including boats and passengers, being temporarily stranded part way up the lift. The first, thought to be in 1979, resulted in a narrowboat with two or three crew on board being marooned for a period of two hours or so until fitters from Northwich Yard, by then referred to as the Repair Yard, managed to free a safety rope which had inexplicably become snagged on the structure. The cause and date of the second is unrecorded but it involved a pleasure boat, with many passengers on board, which was stuck for several hours. Understandably, British Waterways received many letters of complaint.

Two other incidents occurred that same year with respect to the operation of the lift. In the first the Lift Attendant accidentally lowered a gate onto a passing boat. Fortunately, there were no injuries caused and seemingly no significant damage done. The second was more serious and occurred during the Inland Waterway Association's national rally on August Bank Holiday, when one of the caissons was driven too high causing unspecified structural damage.

A very serious mishap, however, occurred on 14 April 1981, which had obvious parallels with the incident of three years earlier. The bare facts are recorded in British Waterways' official inquiry report, which reads as follows:

'On Tuesday, 14th April, 1981 at about 1100, an incident occurred in the operation of the Anderton Lift downstream pontoon [caisson] *when a pleasure craft had entered at the Trent* [& Mersey] *Canal horizon and was about to be lowered to the River Weaver horizon. The pontoon having travelled several feet vertically an incident took place which caused the pontoon to resume its position at the Trent & Mersey Canal level with some force, shaking the narrowboat and occupants.'*

As would be expected, the lift was immediately taken out of service until the cause of the incident had been identified and the extent of any damage assessed. At the time of the occurence, the other caisson was already out of service due to yet another burnt out motor. In all, the lift was closed for a period of seven months. An internal inquiry was set up which took statements from various people, including the Lift Attendant, eyewitnesses and specialist witnesses. The Lift Attendant's account of the incident was that, after having lowered the caisson by 12 inches or so in order to drain the chamber between the aqueduct and caisson gates, the caisson was being lowered further when:

'At about 6ft. down he heard an audible sound of rushing water, the power was switched off and the pawl engaged - he left the control cabin to observe the pontoon [caisson]*. The pontoon, however, was being elevated to the Trent horizon and stopped with a bang and jarr.'*

In the words of Mr. Haigh who was on the boat within the caisson at the time:

*'I brought our hired boat (*Haigh Hall*) onto the Anderton Lift lock* [caisson] *and ran ropes fore and aft over the bollards on the lock wall and held the ends to keep it steady. The lift started to move smoothly downwards, but almost immediately I became aware of considerable vibration, which struck me as being rather strange as I had seen it coming up smoothly earlier in the morning. Then I realised that the rope I was holding became very tight and I had to slacken it rapidly. The front of the boat started to swing around to the left side of the lock and I noticed that the water level was dropping fast. I called to my wife to let go of the rope, which she did, amidst tremendous noise and vibration. Almost immediately after that the entire lock [caisson] came flying upwards a distance that I estimated to be about 4 feet. (Someone who had been watching from another boat told me later he thought it was 6 feet, and also mentioned that he saw some cables fly loose).*

The jolt upwards and sudden stop sent a shower of rust down on the boat and my wife thought she saw something like a brick fly through the air. The boat quickly settled when the lock was back in its up position.

Although things were knocked around inside the boat, no major damage was sustained. The only external signs of damage were some minor scraping, and the tiller knocked out of its seating cup (which was repaired by BWB officials).

No serious injury occurred to any of the 4 of us on board, although my wife bruised her arm and I scraped skin on two fingers.'

The report also records the observations of a British Waterways' employee on arriving at the lift soon after the incident:

'A hire boat, the Haigh Hall *licence No. 67473, owned by Coles Morton, was in the downstream east caisson at canal level ready for lowering to the river . . . All the safety ropes had broken loose from their respective drive wheel mountings on the upper level and were lying on the ground at river level . . . . The crew of the boat (wife and two children) had been assisted out of the boat and made comfortable in the lift cabin. The skipper was asked to do the same and vacated his craft to safety. No one was hurt.'*

Due to the seriousness of the incident with respect to public safety, the newly formed Health and Safety Executive took part in the inquiry. The official findings suggested the following sequence of events:

* Some time prior to the incident, the counterweight of the caisson gate at the aqueduct end had somehow become displaced from its guides.
* At the time of the incident on 14 April the counterweight became trapped, which had the effect of preventing the caisson gate from lowering fully, leaving it open by 5 or 6 inches. This would not have been particularly apparent to the lift operator during the first stage of lowering the caisson to allow the chamber between the aqueduct and caisson gates to be emptied.
* The lift operator was only aware of anything being wrong part way through the lowering of the caisson to the river level when he heard more than the usual rush of water, at which stage he stopped the lift movement and

Photographs of the damage sustained by the boat lift in the accident of 14 April 1981. These appeared in the November 1981 edition of Waterways News, *from which they have been copied, the whereabouts of the original photographs not being known.* ***Top left:*** *The counterweights and safety ropes, which were wrenched from their anchorages, lying on the ground after the accident.* ***Top right:*** *One of the bracing ties, between the guide column and the adjacent 'A' frame, which was fractured in the accident.* ***Bottom left:*** *Close up of the bent handrail caused when the top of the caisson gate frame collided with it during the uncontrolled descent.* ***Bottom right:*** *Close up of the column of an 'A' frame, showing the fracture caused by the force of the accident.* Courtesy The National Waterways Museum, Gloucester

applied the safety pawl in order to lock the lift mechanism in position. The caisson had, by then, descended by nominally six feet.

⋆ The loss of water was so great, however, that within a relatively short period of time the reduction in weight of the caisson was sufficient to cause the counterweights to overcome the frictional forces between the wire ropes and pulleys. The wire ropes therefore began to slide over the fixed pulleys and the caisson started to accelerate upwards.

⋆ At this stage, the safety ropes should have come into operation and performed the function for which they had been specifically designed some 70 years earlier. With slippage of the main ropes over the pulleys the slack in the safety ropes would be taken up and half of the counterweights should have been taken out of the balance. Consequently the caisson should have been brought to rest. Unfortunately, the fixings of the safety ropes to the drive pulleys had not been adequately tightened, seemingly after earlier maintenance works, and all the safety ropes were wrenched from their anchorages and fell to the ground and the runaway caisson continued its upward movement.

⋆ At the top of the lift, the caisson was brought to a sudden halt when the frames of the caisson gates impacted with the machinery platform above. The damage to the caisson and the lift structure suggested that the caisson was out of horizontal alignment during its uncontrolled ascent.

In a letter to British Waterways dated 11 May 1981, the Principal Inspector of Factories for the Health & Safety Executive had the following recommendation with respect to the worn caisson gate counterweight guides, which were the initial cause of the incident:

'Although this is an original design feature of the apparatus it is now clear that unless the guide system is modified to provide full restraint against twisting of the weights a recurrence of the incident cannot be ruled out.'

With respect to the practice of lowering the caissons by a short distance in order to drain the chamber between the aqueduct and caisson gates, a procedure which avoided the Lift Attendant having to leave the control building, the Principal Inspector wrote:

'It is clear that there is room for improved supervision and training of operatives.'

Finally, with regard to the failure of the safety ropes, the Principal Inspector wrote:

'The effectiveness of the existing method of attachment should be reviewed and if necessary it should be replaced by a more secure method.'

Although there were no injuries, the structure was severely damaged in a number of areas. In an article in *Waterways News* in November 1981, under a headline that read '*November Opening for Anderton Lift*', the findings of the structural investigations were succinctly recorded as follows:

'The structural survey revealed a crack at the base of the west outer column in the end frame. Eight steel plate ties were fractured and needed replacing, three bracing panels between columns needed replacing, and a number of joints required bonded repairs. Structural calculations revealed that the remaining sound columns were unbraced and did not provide sufficient support to operate the lift safely, and it was agreed to that the lift had to remain closed until the repairs were effected.

The mechanical report recommended that the design of the rope clamp should be checked and adjusted if necessary; all safety rope anchorages should be dismantled, cleaned and re-assembled to the correct torque loading; new counterweight guides should be installed for aqueduct and caisson gates; the west caisson lifting and safety ropes should be checked to see if they have been stretched beyond their elastic limit at the time of the incident, before deciding whether to renew these ropes and before putting the west lift back into operation; and the caisson gate counterbalance ropes and eyebolts on the west lift aqueduct end, and the caisson gate lifting rope on the west lift river end, should be renewed.'

Two other items of work that were particularly worthy of note, but not referred to in the above extract, were carried out at the same time. The first was the installation of cat-walks along the outer edges of the machinery platform in order to give better access to the safety rope anchorages. Until then, access could be achieved only rather precariously by means of bos'un's chairs, a factor that could help to explain why the safety rope clamps had not been fully tightened. The second, was the repair of the valves that permitted the chambers between the caisson and aqueduct gates to be drained, thereby doing away with the need to lower the caissons by a foot or so to achieve the same effect.

The remedial works were carried out by the Repair Yard and a company called Metalock Limited. In addition, contractors were brought in to shot-blast and paint the machinery platform and support structure at the river end, during which further structural defects were revealed and subsequently repaired. In all, the repair works carried out as a result of the incident amounted to nominally £30,000.

If there were any lessons for British Waterways from this incident they were that:

(a) the lift should be maintained to a standard that is commensurate both with the age of the structure and its use by members of the public.

(b) safe working practices should be set up and followed.

(c) the lift should be manned by adequately trained and experienced personnel

It is reasonable to assume that, in the remaining few years of the life of the boat lift, British Waterways took the

A supporting role for the boat lift, as a backdrop to a 1970s publicity shot for the transport company whose lorry features in the foreground.
Courtesy The National Waterways Museum, Gloucester

A tidy looking Anderton Basin in 1969, quite possibly on a Sunday in view of the lack of acrtivity all round. The coaster Franz Held *sits quietly at the Anderton Depot wharf.* Courtesy The National Waterways Museum, Gloucester

Although traffic through the lift was much reduced, the Anderton Basin remained busy with ships regularly coming up-river to the wharf. In this late 1970s view, the coaster St. Michael, *which often delivered imported shot blast from France, is seen moving out into the channel, prior to swinging round and heading downstream. Work on re-piling the bank around the eastern entrance is in progress.* Courtesy Cyril Bailey

appropriate steps to address these.

★★★★★★★★★★

Sadly, within two years of it being re-opened after the incident in April 1981, the final and inevitable decision to close the Anderton Boat Lift was taken. By then the original elements of the boat lift were between 75 and 110 years old. It was therefore not particularly surprising that the ravages of time and the corrosive conditions at Anderton finally resulted in the lift being declared as structurally unsound and not safe for use by the public. The events leading up to the closure were recalled by Brian Haskins in the following way:

'It closed as a result of painting taking place by means of a contract let by our Leeds office (then Technical Services but formerly the Regional Engineer). The contract painters came in and when they applied the shot-blasting they just shot holes in the structure and they said we can't paint this! So we got two or three or four of the [Repair Yard] *welders to weld plates or weld repair patches where the holes had been. The welders realised at the end of a couple of weeks that they couldn't keep pace with it, so we got contract welders in to help them. I think at the time there were six contract welders working ahead of the painters or just behind the shot-blasters, but they found they couldn't keep pace with it even with that. At that point the Manager of Technical Services at Leeds, Mr. Ghulam Haider, stepped in and decided that a full scale survey was required to assess the condition not only of the Aqueduct, but the whole lift structure as well. These surveys, carried out by consultants, confirmed his view that the structure was very badly corroded and that some elements were in a parlous state.'*

And so in Autumn 1983, after over 100 years of almost continuous operation, the Anderton Boat Lift was finally taken out of service.

Commemorative plate by Wedgewood for the centenary of the Anderton Boat Lift in 1975.
Courtesy The Boat Museum, Ellesmere Port

INFORMATION SOURCES FOR CHAPTER 9

Steve Astles – *Engineering Supervisor on the Weaver Navigation for British Waterways, North West Region.*
Arthur Dodd – *Labourer, Plant Operator and Dredger Captain for the Weaver Navigation and subsequently British Waterways, North West Region from 1938-1979. Interviewed on 11 September 1999 at Northwich.*
Brian Haskins – *Area Engineer at Northwich for British Waterways, North West Region from 1968-1988. Interviewed at Northwich on 3 September 1999 and subsequent discussions.*
Jack Horton – *Joiner and Foreman for the Weaver Navigation and subsequently British Waterways, North West Region from 1933-1975. Interviewed on 2 September 1999 at Northwich.*
Frank Westwood – *Repair Yard Manager at Northwich for British Waterways from 1964-1981. Interviewed at Northwich on 3 September 1999 and subsequent discussions.*

The Inland Waterways of England, *by L.T.C. Rolt. Allen & Unwin, 1950.*
Anderton for Orders – Memoirs of a Working Boatman in the early 1950s, *by Tom Foxon. J. M. Pearson & Associates, 1988.*
The Illustrated History of Canal and River Navigations, *by Edward Paget-Tomlinson. Sheffield Academic Press, 1994.*
Anderton Barge Lift, Northwich. *Report August 1965 by Husband & Co.*
Report on the Anderton Boat Lift - Northwich. *February 1972 by Husband & Co.*
Anderton Lift Inquiry - Incident at the Anderton Lift on 14 April 1981. *Report by Principal Engineer (North), July 1981.*
Anderton Barge Lift Rehabilitation - Presentation of Alternative Outline Schemes. *Wallace Evans & Partners, 23 May 1984.*

A 1980s interior view, looking down on the drained and empty caissons showing their plated and rivetted construction. There is much evidence of corrosion, particularly on the top surfaces. Whilst this picture was one of a series commissioned to show the condition of the lift, it also demonstrates the attractions of the structure as far as photographers are concerned; all angles, patterns of construction and the effects of light filtering through the beams. Photograph Lawrence Sands, courtesy The National Waterways Museum, Gloucester

CHAPTER 10
THE STATIC MONUMENT

'It's time it was put together; it's only just an ordinary job! . . . If Saner or Marsh had been in charge, they would have done it by now. They wouldn't have tolerated all this talk and wanting money and God knows what. They would have got on with it . . . Once they started they would have found the money, wouldn't they!'
Extract from interview with Jack Horton on 2 September 1999

If only it was that simple. Who knows? Perhaps Jack is right about Saner and Marsh but I doubt it. Surely the telling factor is that in their lifetimes the Anderton Boat Lift was a vital asset to the Weaver Navigation upon which much of the income to the Trust was dependent? It would have been financially unacceptable to the Trustees to permit the boat lift to be closed for an extended period of time. As was the case in 1908, the money to effect any necessary repairs would have been found by some means or other. However, the circumstances today are altogether different. Even if the money was readily available, the boat lift is no longer viable in terms of income from traffic alone and other means of justifying the expenditure for its restoration have to be identified and explained. Furthermore, faithful restoration of such an old structure, especially one that has been scheduled as an Ancient Monument, is fraught with difficulties. The project is by no means straightforward and is further complicated by the need to apply modern-day health and safety standards to what is basically a Victorian structure.

★ ★ ★ ★ ★ ★ ★ ★ ★ ★

British Waterways lost little time in addressing the issues. Not long after closure of the lift in 1983, consulting engineers Wallace Evans & Partners had been appointed to investigate alternative schemes for refurbishment. Their brief, initially at least, consisted of the appraisal of four basic options for the lift structure, as follows:

★ Restoration of the lift to its full operational condition with two caissons.
★ Partial restoration of the lift to achieve single-caisson operation.
★ Modification of the lift to modern synchro-lift type of operation with refurbishment of the necessary elements of the lift structure only.
★ Preservation of the lift as a static monument.

The third of these options was apparently to be investigated by an organisation other than Wallace Evans. Consequently, there is little documentary evidence available as to what the proposed 'modern synchro-lift system' entailed.

The investigation by Wallace Evans involved a structural assessment of the lift in its badly corroded condition, with the identification of the relevant strength parameters for design purposes and assessment of the carrying capacity of the corroded elements. The individual elements of the structure included in the analysis were the drive machinery, including the pulleys, gearwheels, drive shafts and ropes; the access structures, including the walkways and handrails; the machinery floor joists and supporting beams; the 'A' frames including the vertical columns, the angled columns and the base plates; the horizontal and diagonal bracing between the 'A' frames; the caissons; the 1875 guide columns and the masonry dry well. (It should be noted that, at this stage, the investigations concentrated on the lift structure and did not take account of the condition of the aqueduct and ancillary structures).

In their report of May 1984, Wallace Evans & Partners confirmed the very poor condition of the boat lift structure as follows:

'The structural survey has identified the following elements which are sufficiently corroded to give cause for concern for the integrity of the Boat Lift structure:

(i) North East cross drive shaft support beams
(ii) Machinery floor joists
(iii) Longitudinal beams, especially the southern end
(iv) Three internal cross beams
(v) Columns and raker columns, particularly those at the north end
(vi) Caisson wall and floor plates'

Their report went on to advise that:

'In general, even after measures have been taken to restore local stability and integrity of connections, the loss of metal as a result of corrosion has so reduced the potential load carrying capacity of the structure that one of the two following strategies must be

Scrap metal being unloaded onto the quayside at the Anderton Depot about 1980, with the boat lift looming behind.
Photograph by Lawrence Sands, courtesy Cyril Bailey

adopted to restore the lift to a safe working condition:
either the structural inadequacies must be made good by increasing the strengths of existing members or by replacing them completely;
or the forces in the members must be reduced, by reducing applied loads, so that they are within the residual capacities of the members.'

The consultants identified several variations on the four basic options, one of which was the refurbishment of the lift structure with reduced capacity caissons. After detailed assessment and comparisons, including structural calculations, cost estimates and programming, the consultants cautiously recommended the full refurbishment of the structure, with strengthening of several of the main machinery deck girders by supplementary steel trusses and the filling of the 'A' frame columns with mass concrete. The estimated cost of this option was £215,000.

Not long after, Wallace Evans & Partners were asked to extend their investigations to include three other options. The first was a variation of the original full refurbishment option which involved the replacement of the tubular steel 'A' frames, rather than the filling of the existing ones with concrete, together with replacement of the badly corroded main girders to the machinery platform. The second new option involved a similar variation to the original partial single caisson option, whilst the third option was a cheaper alternative to the original abandonment option. It was described in the following way:

'Minimum cost abandonment based on completely removing the loads of the caissons and counter-balance weights from the structure. The minimum amount of steelwork repairs will be carried out to ensure the Lift's short term stability. Frequent inspection and maintenance in the future.'

Wallace Evans costed the latter at £47,236 and it was this option, or works very similar to it, that British Waterways were eventually to adopt in 1987.

★ ★ ★ ★ ★ ★ ★ ★ ★ ★

Further assessments and investigations of the corroded structure followed and, before long, detailed structural analysis revealed that the lift was in worse condition than originally thought. It was consequently decided to minimise the loading on the weakened 1908 'A' frames by removing the cast iron headgear and electrical equipment from the machinery platform, and lowering the caissons and counterweights to the ground. Outside contractors, M. E. Mechanical Handling Ltd were appointed for the work, with Project Management Partnership acting as British Waterways' representatives on site. The works began in November 1987 and took about six weeks to complete.

Right: *Scaffolding alonside the aqueduct in the early 1980s. Cleaning and painting of the metalwork is in progress.*
Below right: *The gear wheels and pulleys being sprayed with a protective lubricant, again in the early 1980s.* Both photographs Lawrence Sands, courtesy The National Waterways Museum, Gloucester

The majority of the items removed from the machinery platform were stockpiled on the ground to the east of the lift, whilst the remainder was taken to the Repair Yard at Northwich for safekeeping. Sadly, it is in this denuded condition that the structure can be seen in today.

In addition to supervising the dismantling of the headgear and lowering of the caissons and counterweights, Project Management Partnership also carried out detailed condition inspections and reports on each of the mechanical and electrical components being removed, supported by photographs as appropriate. Non-destructive testing was undertaken at the same time. The consultants' extensive report consisted of four separate volumes.

★ ★ ★ ★ ★ ★ ★ ★ ★ ★

By February 1989, British Waterways had produced a tender document for the total replacement of the superstructure of the electrical boat lift and, one assumes, were considering inviting outside contractors to tender for the work. The description of the proposed work contained in the tender document is as below:

'The existing metal boat lift structure is to be completely dismantled and replaced with a new one to replicate the original so as to be fully in keeping with its status as an Ancient Monument. The existing structure is to be carefully measured in every detail and a complete and comprehensive set of drawings are to be produced so that each replacement member and hence the whole structure can be accurately reproduced to its existing form. The two caissons currently supported on stocks (within the dry well) will remain in the reception dock during the whole of the works. The reception dock is to be stanked off from the Weaver Navigation, any water in it is to be pumped out and the whole to be kept dry by pumping during the course of the works.

The stonework surrounding and embedding the earlier built portion of the structure is to be carefully taken down in the dry and rebuilt after the replacement structure has been completed. Temporary support is to be provided to the central girder of the aqueduct whilst that portion of the earlier built structure is being replaced. The water tight guillotine gates are to be reinstalled at the River Weaver Navigation end of the structure before the stanking is removed. The whole site is to be left clean and tidy on completion of the works.'

It is apparent that a strictly like-for-like replacement was not intended as there is a clause later on in the tender document entitled 'Specified Alterations', which details five separate areas in which enhancements to the structure were considered necessary. These included increased thicknesses to the tubular steel elements of the 'A' frames, modification of the spherical bearings between the 'A' frames and the machinery platform so they would be fixed against rotation, and placing concrete within the tubular cast iron columns of the 1875 structure. A particular point of interest is the number of clauses in the document devoted to surface preparation and painting of the new structure; a recognition of the problem that bedevilled the lift since it was first built and which, to a very great extent, still exists.

The proposed works detailed in the tender document would presumably have been the first stage in returning the electrical boat lift to working condition; the refurbishment of the caissons was not part of the proposed works nor is there any mention of the mechanical and electrical equipment.

These initial proposals for restoration were not progressed, however, principally, it seems, due to the involvement of English Heritage, who had jurisdiction over the boat lift as a Scheduled Ancient Monument. From this point on, the vexed question of restoration rather than replacement came into the frame alongside the structural integrity of the lift. Much of the ensuing engineering investigations consequently gave consideration to the percentage of the original 1875 and 1908 materials or structural elements that could be retained with each solution for refurbishment.

★ ★ ★ ★ ★ ★ ★ ★ ★ ★

In June 1989, Cheshire County Council carried out an archaeological survey of the boat lift site in order to identify any hidden historical features. The investigations consisted of surveys of structural remains in the vicinity of the lift and basin, together with documental research at the Ellesmere Port Boat Museum and the County Records Office at Chester.

The works on site, which involved exploratory excavations by volunteer labour, concentrated on two specific areas; the sloping ground between the existing river basin and the canal, and the area of the backfilled basin to the east of the lift. The investigations revealed remains of the following items:

★ Salt chutes – footings of a chute were found to the west of the lift.

★ Loading platforms – a stone wall at the canal level which was once part of the loading platform to the chute.

★ Engine house – remnants of a brick-built engine house on the sloping ground adjacent to the central chute. The engine house was thought to be mid-19th century and was believed to have powered a steam hoist to raise goods from the river to the canal.

★ Dockside structures – remains of walls and buildings.

★ Trackway – a track running diagonally down the slope from the canal basin to the northwest corner of the river basin. Part of the track comprised squared stone setts whilst the remainder was made up from sandstone rubble. The track was thought to have been used to take horses between the canal and stables that were at the side of the river basin.

★ Dockside steps – the remains of a sandstone retaining

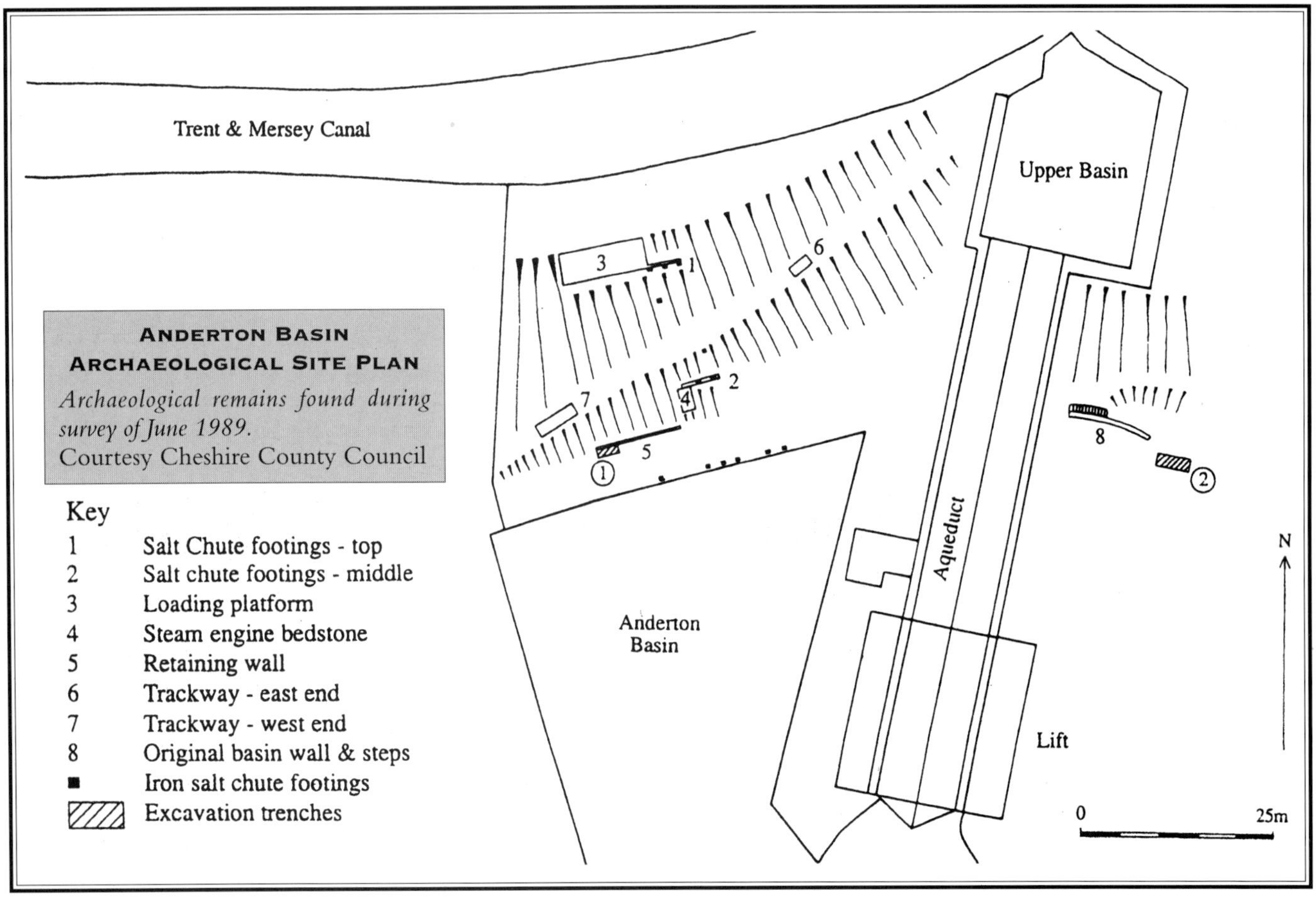

ANDERTON BASIN ARCHAEOLOGICAL SITE PLAN

Archaeological remains found during survey of June 1989.

Courtesy Cheshire County Council

Removal of the wheels, counterweights and pulleys took place over a six-week period in November and December 1987, with the aid of a mobile crane with a telescopic jib, to relieve stress on the structure. Left, courtesy The National Waterways Museum, ; right, Edward Paget-Tomlinson

wall together with thirteen well-worn steps were found to the east of the aqueduct. It was suggested that these were once part of the pathway leading between the canal basin and the, now backfilled, eastern river basin.

The archaeological report concluded with the recommendation that further investigations should be carried out as part of any development of the boat lift site.

★ ★ ★ ★ ★ ★ ★ ★ ★ ★

During 1991 and 1992, British Waterways carried out extensive structural surveys and analyses of the lift structure. One of the objectives was to ascertain how much of the original structure could be retained if the lift was to be restored and, consequently, the extent of the replacement works necessary. Their report, which was dated March 1992, consisted of six separate volumes. The study concentrated on two basic options:

Option 1. Full restoration of the 1908 electrical lift with both caissons operational.

Option 2. Restoration with just one, smaller, caisson in operation.

Further non-destructive testing and studies were carried out on the lift structure, (not including the aqueduct), and each element was allocated a colour coding. 'Red' indicated that the element had deteriorated beyond repair and would have to be replaced; 'Amber' that the element could be incorporated in the restored structure but only after repair works; and 'Green' that the element was in sufficiently good condition to be incorporated within the restored structure with little or no repair works. The report advised that, on the basis of the combined length of all the structural elements, over half of the lift structure would need to be replaced for Option 1, whilst less than one third of it could be retained without further work. The situation for Option 2 was that one eighth needed replacement and slightly less than half required no remedial work. (These codings are still visible on the structure today in the form of coloured bands painted around the various elements).

A paper by British Waterways dated October 1992 and entitled *The Re-activation of the Anderton Boat Lift – Trent and Mersey Canal – Weaver Navigation,* identified the likely cost of the first of the above options, full restoration, as £2.8 million and the cost of restoration with a single, reduced size caisson as £2.3 million. It should be noted, however, that these estimates are thought to relate only to the lift structure and, as such, did not allow for any remedial works that would be needed to the aqueduct or the canal and river basins.

In 1993 a series of tests was carried out on site to ascertain, for design purposes, the coefficient of friction between the wire lifting ropes and the cast iron pulleys. Later that year, Parkman Consulting Engineers of Salford were commissioned by British Waterways to undertake an audit of all the investigations, studies and analyses carried out up to that date and to make recommendations for the way forward. The findings in their report dated March 1994 are summarised as follows:

★ English Heritage and the Health & Safety Executive would both need to be consulted with respect to any proposals for restoration.

★ The restoration works would require significant capital outlay. (Parkman confirmed the earlier estimates for the two restoration options and identified the likely costs of options relating to retaining the lift as a static monument).

★ Notwithstanding the amount of investigation works

IMAGES OF THE BOAT LIFT IN 1999 – AWAITING RESTORATION

Photographs by Martin Grant, Lewin, Fryer & Partners, and author

The aqueduct and control cabin looked in tidy condition, the boat lift less so, when photographed in 1999. Photograph Martin Grant

carried out by that time, there were several areas which warranted further investigations. These included condition surveys of the aqueduct, the canal basin, the river basin and the plant building beneath the aqueduct.

Possibly as a consequence of the first of the above recommendations from Parkman, British Waterways formally applied to English Heritage for their approval to the full restoration of the electrical lift on 6 July 1994 and consent for the works was duly granted on 3 October of the same year. The consent, however, contained six conditions, four of which were of particular interest:

'iii. original metalwork shall be re-used wherever possible;
iv. any replacement metalwork shall be of a colour and finish which matches the existing;
v. equipment and machinery shall not be used or operated in the scheduled area in conditions or in a manner likely to result in damage to the monument or disturbance other than necessitated by the authorised works and carried out in accordance with the agreed specification;
vi. photographs and plans and full engineering drawings shall be taken of the monument before the start and after the completion of the works and a set of prints and drawings shall be sent to the Historic Buildings and Monuments Commission at the address shown . . .'

★★★★★★★★★★

Consulting engineers, Lewin, Fryer & Partners, based at Hampton in Middlesex, first became involved in the project in November 1994. They were, at that stage, commissioned by British Waterways to review the feasibility of the proposed restoration works, to carry out a detailed technical appraisal and prepare tender documents for the full restoration of the electrical lift. Around that time, detailed investigations into aspects relevant to restoration of the boat lift were carried out, either by the consultants themselves or by specialist outside organisations under the supervision of Lewin, Fryer & Partners. The primary objective of these was to fill in the gaps of knowledge with respect to the condition of all the elements of the structure, not just those that were readily accessible. To this end, the further investigations included:

★ **Soil sampling and testing** to gain a better understanding of the underlying geological strata, with particular reference to the load bearing capabilities and the extent of contamination by salt and other contaminants.

★ **Silt sampling and testing** from the canal and river basins to identify any contamination that could warrant special disposal methods.

★ **Foundation investigations** to assess the extent and condition of the masonry elements of the lift that are below ground.

★ **Metallurgical investigations** to identify the metals used in the various parts of the lift and aqueduct structures, and to obtain appropriate strength parameters for design purposes.

★ **Topographical surveys** to provide detailed contoured plans of the boat lift site and to evaluate the verticality of the structural components.

★ **Non-destructive testing on the aqueduct** to establish the extent of the corrosion and to identify the thickness of sound metal remaining on the main structural elements. The opportunity was also taken to confirm previous tests carried out on the lift structure.

★ **Detailed examination of the electrical components** to assess their present conditions and to identify the works needed to return them to full working order.

★ **Review of the condition of the mechanical components** for the same reasons as above.

★ **Weighing of the caissons and counterweights** to determine the precise weights of the two caissons and the 72 counterweights.

★ **Identification and assessment of the health and safety implications of restoration of the boat lift** to ensure modern health and safety guidelines are followed.

★ **Condition surveys of the canal and river basins** to identify any structural defects and to assess the extent of siltation.

★ **Building surveys** to assess the structural condition of the ancillary structures, including the wooden control building on the lift structure, the brick 'engine house' building beneath the aqueduct and the brick river gate control buildings.

The report by Lewin, Fryer & Partners, entitled *Detailed Technical Appraisal for the Restoration of the Anderton Boat Lift*, concluded:

'From the preliminary design the extent of restoration works has been determined and costs estimated for four principal options. The estimated whole life cost for the recommended option for (full) restoration, operation and maintenance is £4.65m. The Consultant has also made recommendations to improve the operation of the lift, to make it safer and to comply with current day working practices and legislation. The estimated whole life cost of the Monument, for the recommended option, with all the recommended improvements is £5.6m. (For comparison purposes, the capital cost of the full restoration works was estimated as £3.51 million) These costs will be re-evaluated during detailed design when the extent of the works will be more accurately defined . . . . It has been determined that it is possible to restore the Monument to full working order to reproduce the 1908 design philosophy, operation and general appearance. The Monument would be restored to the condition of 1976 when the lift structure was scheduled as an Ancient Monument.'

During the course of the further investigations summarised above, consultations took place with English Heritage and the Health & Safety Executive. The proposals for the full restoration of the boat lift were discussed and general consent to the works was obtained. It should be noted that the estimated cost of the refurbishment works by then included remedial works to the aqueduct and ancillary structures.

★★★★★★★★★★

British Waterways' original intention was to restore the boat lift to electrical operation, generally to its condition after conversion to electricity by J. A. Saner in 1908. However, in 1997, the decision was taken by British Waterways, in consultation with English Heritage, to restore the lift to hydraulic operation, generally as conceived by Edwin Clark in 1875, with the 1908 structure being retained as a static monument.

This significant change is reported to have occurred after an application for a grant from the Heritage Lottery Fund in January 1997 was unsuccessful. The bid amounted to £11 million and was for the full restoration of the 1908 electrical boat lift, the construction of an extensive visitor centre and the provision of an endowment fund for future maintenance and repairs. On consideration of the restoration proposals prior to re-submission of the bid, it was realised that an unobtrusive way of providing a secondary braking system, in the event of failure of the electrical lift, would be to install a hydraulic ram to act as a cushion beneath each of the caissons. (Coincidentally, this proposal had parallels with a suggestion made by Mr. Bramwell at a meeting at the Institution of Civil Engineers in London in March 1876 – see Chapter 6). From this position it was a relatively short step to restoration of the lift to hydraulic operation and to retain the 1908 structure as a static monument. Another benefit of this option, as far as English Heritage was concerned, was a reduction in the extent of the replacement, rather than restoration, works needed.

This decision by British Waterways, however, necessitated further investigations and studies in order to identify the condition of those elements of the 1875 hydraulic boat lift structure that, hitherto, had been neglected. This applied particularly to the ram shafts and access tunnels beneath the floor of the lift basin. Consequently, in February 1998, Lewin, Fryer & Partners were commissioned to extend their studies accordingly. Their work included a re-assessment of the 1908 structural members with significantly less loading than previously, investigation of the ram shafts and interconnecting chambers, and identification and assessment of methods of restoring the boat lift to hydraulic operation. They were also asked to give consideration to possible measures for public access onto the restored boat lift. The main findings, which are contained in the consultants' report dated April 1998, were:

★ On the basis that the 1908 lift structure remains as a static monument, the majority of the 1908 structural elements could be retained, albeit with strengthening works to achieve a design life of 99 years, as required by British Waterways.

★ The concrete counterfort foundations of the 'A' frames to the 1908 lift structure would need to be replaced to achieve the 99 year design life.

★ The 1875 hydraulic rams, control valve and connecting pipework are no longer on site and the ram shafts and

Goole Star *departs Anderton Basin in the 1980s. On the left is the* Grimsby *which, from the angle of her bow in the water, is partially unloaded.*
Courtesy The National Waterway Museum, Gloucester

A viewpoint which visitors will be able to enjoy again before too long; the Anderton Basin and the industrial surroundings of the River Weaver from off the top of the lift. The coaster Grit, *of London, is unloading coal at the Anderton Depot in this late 1980s photograph.*
Courtesy The National Waterway Museum, Gloucester

associated chambers appear to be in reasonably good condition, but require further clearance of silt and debris.

★ After consideration of several options, it was decided that a single, centrally mounted, single-acting plunger ram under each caisson, as originally designed by Edwin Clark, was the preferred arrangement.

★ Unsupervised access for members of the public onto the restored structure could be permitted to safer areas of the elevated boat lift structure, with guided tours to the other parts.

★ The estimated cost of the full restoration works to the lift, including the aqueduct and other ancillary structures, was £5.5 million.

In addition to the technical aspects of restoring the boat lift, various proposals have been considered for the development of the boat lift site so that, in conjunction with other attractions in and around Northwich (i.e. the Salt Museum, the Lion Salt Works and the Ellesmere Port Boat Museum) more visitors will be attracted to the area. Such considerations have included better access facilities to the site both by road and boat; circular walkways around the site; and a purpose-built visitors' centre.

★ ★ ★ ★ ★ ★ ★ ★ ★ ★

After the closure of the lift in 1983, several organisations have been established with the aim of actively supporting and encouraging British Waterways during their investigations prior to restoration. These organisations included:

The Anderton Boat Lift Development Group: Formed in 1986 with the intention of keeping the issue of restoration of the boat lift in the public arena. The Group was a non-political, independent consortium of Cheshire County Council, Northwich Town Council, The Inland Waterways Association, The Trent & Mersey Canal Society and Vale Royal Borough Council.

The Anderton Boat Lift Trust: Established in 1994 with representation from British Waterways, Cheshire County Council, The Inland Waterways Association, Mid Cheshire Chamber of Commerce, Trade & Industry, The Trent & Mersey Canal Society, Vale Royal Borough Council and other interested parties. It was formed primarily in order to raise funds towards the restoration of the boat lift.

Friends of the Anderton Boat Lift: An organisation set up in 1996 by enthusiastic and interested individuals keen to see the restoration and operation of the boat lift. The Friends work alongside the Anderton Boat Lift Trust and have set up a temporary visitor centre at Anderton which is manned by volunteers.

Perhaps the best way to explain the activities of the Anderton Boat Lift Trust and the Friends of the Anderton Boat Lift is to quote verbatim their shared vision statement for the boat lift:

'Anderton Boat Lift will be restored as an operational link between the Weaver Navigation and the Trent & Mersey Canal. It will be internationally important as a heritage site and will be the culmination of the many partners' aspirations to provide economic, social, environmental and recreational benefits to all. It will provide a sustainable leisure and educational focus for Northwich and its surrounding area for future generations.'

★ ★ ★ ★ ★ ★ ★ ★ ★ ★

And so, seventeen years or so after the enforced closure of the Anderton Boat Lift in 1983, all the pieces of the jigsaw are seemingly in place and the final preparations for the restoration of the hydraulic boat lift to full working order can be started.

INFORMATION SOURCES FOR CHAPTER 10

Jack Horton – *Employee of the Weaver Navigation and British Waterways, North West Region 1933 -1975. Finished as Senior Area Inspector, British Waterways. Interviewed on 2 September 1999 at Northwich.*

Anderton Barge Lift Rehabilitation - Presentation of Alternative Outline Schemes. *Report by Wallace Evans & Partners, Consulting Engineers, 23 May 1984.*

Reconstruction of Anderton Boat Lift Linking the Weaver Navigation (Commercial) with the Trent & Mersey Canal (Cruising), *Report by Technical Services of British Waterways at Leeds (formerly Regional Engineering), February 1989.*

Anderton Boat Lift - Reports and Studies with Cost Summary and Assessment of Time Duration for Works Necessary for Reconstruction, *Report in 4 vols. produced by the British Waterways Board, Engineering Department in conjunction with Project Management Partnership, 1988.*

Anderton Basin – An Archaeological Survey, *Cheshire County Council, October 1989.*

The Re-activation of the Anderton Boat Lift – Trent & Mersey Canal – Weaver Navigation, *British Waterways, 7 October 1992.*

Anderton Lift – Proposals for Restoration – Technical and Financial Audit, *Parkman Consulting Engineers, March 1994.*

Detailed Technical Appraisal for the Restoration of the Anderton Boat Lift, *Lewin, Fryer & Partners, October 1996.*

Anderton Boat Lift Restoration - Study for 1875 Restoration and 1908 Static Monument, *Lewin, Fryer & Partners, April 1998.*

Resubmission of Heritage Lottery Fund Bid for the Anderton Boat Lift, *British Waterways & Anderton Boat Lift Trust, April 1998.*

Gear wheels, aqueduct, control cabin and lift, 1999. Author

CHAPTER 11
THE LIFT IN THE THIRD MILLENNIUM

'£3M GRANT TO RESTORE GIANT BOAT LIFT – The Anderton Boat Lift, one of the Victorian's most magnificent and eccentric feats of industrial engineering, is to be restored to its original working order . . . . When it was built in 1875 by Edwin Clark, it was the first of its kind anywhere in the world and revered as a testimony of the ingenuity and sheer daring of 19th century engineers. British Waterways, which describes the lift as a cathedral of the Industrial Revolution, yesterday announced a National Lottery grant of £3.3 million towards the £7 million restoration costs.'

Extracts from an article in The Times, *31 March 1999*

Sixteen years after the enforced closure of the Anderton Boat Lift due to its poor structural condition, this article in *The Times* newspaper in March 1999 announced the decision by the Heritage Lottery Fund to support an application for funding towards its full restoration. This successful request was for the restoration of the 1875 hydraulic boat lift; the previous application, almost a year earlier had been to restore the 1908 electrical lift. The applications were made by British Waterways in partnership with the Anderton Boat Lift Trust.

The article in *The Times* contains some very effusive quotes from British Waterways' officers. The first is by Derek Cochrane, Regional Director of British Waterways - North West Region:

'There is no doubt that the Anderton Boat Lift is a monument of international significance. When built it was the biggest step in water transport since the building of the Pontcysyllte Aqueduct 70 years earlier. Later boat lifts in Belgium and France copied Anderton but its design was never bettered.'

Phil Parker, who at that time was Project Manager for British Waterways - North West Region, was quoted as follows:

'I feel deep envy towards those who built the original lift because they took such a risk. It was highly entrepreneurial and one hell of a leap for the technology.'

★ ★ ★ ★ ★ ★ ★ ★ ★ ★

The money promised by the Heritage Lottery Fund for the restoration works was slightly less than half the full amount of the £7 million required for the project, £4.8 million of which is for the restoration of the lift structure itself. The best way of setting out British Waterways' intentions for restoration of the boat lift and development of the site is by quoting from the Heritage Lottery Fund bid document, dated April 1998:

'This resubmission to the Heritage Lottery Fund is the culmination of many years work by British Waterways and other partners including many hours provided by volunteers. British Waterways and the Anderton Boat Lift Trust have raised a considerable amount of funding from English Heritage, the Vale Royal Borough Council, volunteers and private companies. This is a result of the wide recognition of the Boat Lift as a structure of great heritage value as well as a connection between two navigations.'

Further on it summarises the proposals for the boat lift and the site:

'The Project consists of the following elements:
1) The repair and maintenance to full working order of the Boat Lift using hydraulics as the method of lifting but retaining the refurbished 1908 modification in situ ensuring:
★ Enhancement and preservation of the heritage value of the structure and site.
★ Navigation between the Weaver Navigation and the Trent & Mersey Canal.
★ High quality engineering to up to date standards and codes.
★ Highest possible operational safety standards for safe use of the Anderton Boat Lift.
2) Provision of high quality, entertaining and educational interpretation of the national heritage site ensuring:
★ Awareness of the heritage value of the Boat Lift by visitors and local residents.
★ Sustainable development of the Boat Lift to ensure maintenance and operation of the structure.
★ Creation of strategic links with complementary industrial heritage and waterway tourism sites.
3) Creation of supporting links to the community and environment via:
★ Landscape interaction with the adjacent Anderton Nature Park, Mersey Forest, Brunner Mond and waterways including visitor facilities for pedestrians, boaters and car-borne visitors.
★ Apprenticeship training in key skills such as boilermaking and shipwrights.
★ Local community involvement in the restoration, operation, use and interpretation of the Anderton Boat Lift.
★ Sustainable economic, social and environmental benefits for the surrounding area through the increase of tourism, leisure and educational activity.'

The bid document states that the new hydraulic rams will be similar to those designed by Edwin Clark for the 1875 lift and will be installed within the original cast iron shafts which are still in place below the dry well of the boat lift. The new rams will, however, make use of modern, corrosion resistant materials. The caissons, the guide columns and the associated support structure of the 1875 lift, together with the aqueduct, is to be refurbished retaining as much of the existing metalwork as is practicable. The 1908 structure, as designed and installed by Col. J.A Saner, which includes the 'A' frames and the machinery deck, is to be dismantled to enable full restoration to be carried out before it is reinstated, as a static monument, over the refurbished hydraulic lift. The objective is to carry out the restoration works to a high standard so as to achieve a design life for the structure of 99 years. The operation of the restored lift is to be as simple as possible whilst incorporating safety features that are commensurate with modern-day requirements.

As well as describing the proposed works, the bid document provides details of the means of funding the operation and maintenance of the restored boat lift. It predicts that the re-opening of the Anderton Boat Lift will attract some 200,000 paying visitors to the site each year, of which nominally 50,000 will make use of the proposed trip boat on the lift. The currently intended charge for the use of the lift by private or hire boats is a toll of £27.50 per boat for a return trip, an amount that might also include entry to the site and a guided tour of the structure.

* * * * * * * * * *

The above extracts from the bid document contain little detailed information regarding the restoration of the boat lift structure itself. Technical details, however, were contained within the report by Lewin, Fryer & Partners entitled *Anderton Boat Lift Restoration – Study for 1875 Restoration and 1908 Static Monument*, dated April 1998 and it is to this document that I now turn for a fuller explanation of the restoration proposals:

A – 1908 Lift Structure: *Dismantled with the majority of parts refurbished rather than replaced. The headgear is to be refurbished and installed in working order with a token sample of the counterweights replaced. Those elements which are missing or are beyond economical repair are to be replaced . . . . The 1908 vertical and bracing columns ('A' frames) can all be retained, albeit with repairs. There is some concern with regard to the longevity over the 99 year (design) period. Periodic inspections and a high level of maintenance would be required to ensure no further significant deterioration. In addition, several minor bracing members will require replacement . . . The 1908 foundations and supporting dry well walls can be utilised without major works. They will, however, need regular inspections to ensure there is no further significant deterioration of the concrete and masonry elements.*

B – Aqueduct: *Refurbished in situ, with elements that do not have sufficient thickness to meet structural and full corrosion requirements replaced or refurbished, depending upon the extent of work required . . . . The control building is to be repaired along with the support assembly and longitudinal floor beams.*

C – 1875 Guide Frame Structure and Hydraulic Lift: *The guide frame columns are in good order and require little more than cosmetic works. The bracing members, however, require refurbishment depending upon current thicknesses . . . . The river gate platform is to be dismantled and refurbished prior to reconstruction . . . . The river gates themselves are to be repaired to ensure sufficient structural capacity and corrosion allowance for the full design life.*

D – Caissons and Caisson Gates: *Structural alterations are required to the caissons to facilitate the ram connection. Existing members are to be curtailed and some replaced to allow the caisson support assembly to be installed . . . . The caisson gates and frames are to be refurbished with new ropes and seals.*

E – Dry Well and Lift Foundations: *The concrete base to the dry well to be made good with a new screed applied. A permanent pump and outfall installed to dewater the dry well during operation of the lift . . . . The existing ram shafts require the remaining silt and debris to be cleared to allow the new ram system to be installed. The temporary caisson support frames are to be dismantled and disposed of off site . . . . It is likely that the existing concrete around the ram shafts would need to be replaced with high load bearing foundations.*

F – River Basin: *The cofferdam is to be de-silted and removed. The estimate does not allow for further dredging works in the river basin . . . . The river basin walls are to be repaired for both cosmetic and safety reasons.*

G – Canal Basin: *As per the river basin but with the addition of a bentonite lining to the basin base and a cut-off wall installed between the canal and the basin.*

H – Buildings: *The buildings are to be refurbished to a high standard to allow them to be utilised to house part of the ram system or to provide information display areas.*

I – Miscellaneous: *Upgrade/re-new the electrical supply and distribution system. Lighting to control room and buildings to be replaced.*

* * * * * * * * * *

So, about 125 years after the construction of the hydraulic lift and about 90 years after the conversion of the boat lift to electrical operation, the many and complex pieces of the jigsaw to enable the Anderton Boat Lift to be restored to working condition appear to be in place. It is to be returned to hydraulic operation, much as it was when originally designed by Edwin Clark and built by Emmerson Murgatroyd & Co. Ltd. in 1875. The structural elements which were added for the electrical conversion works in 1908, as designed by Col. J. A Saner and constructed by the workforce of the Weaver Navigation, are to be restored and retained on site as a static monument. It is currently proposed that some of the 1908 drive cogs and pulleys will be reinstated on the machinery platform to demonstrate the workings of the electrical lift.

With hindsight, it is evident that both the original hydraulic lift and the converted electric structure should be acknowledged as engineering masterpieces worthy of retention for future generations to marvel at. I personally have no doubt that the Anderton Boat Lift should be

Computer generated images of the proposed restored boat lift and new visitors' centre. Courtesy Buttress, Fuller, Alsop & Williams

THE ANDERTON BOAT LIFT IN THE 21ST CENTURY

restored and I hope that those who once did, such as Mr. G. Davies whose letter I quoted at the beginning of Chapter 1, may be won over to the merits of restoration.

The original hydraulic structure was a daring response by the Trustees of the Weaver Navigation to commercial pressures for an improved connection between the Trent & Mersey Canal and the Weaver Navigation; the time and cost of overland transshipment of goods between the canal and river, especially at a time when waterway traffic was so heavy, was driving business away from the navigation and onto the canal.

The operating mechanism of the hydraulic lift, although not totally unique, was beautifully simple and efficient. With the load of each caisson being carried by a single vertical ram, the remainder of the superstructure was relatively lightweight and has been described as elegant and graceful.

The boldness of the Trustees was more than repaid by thirty years or so of operation of the hydraulic lift, during which time it conveyed over 3 million tons of cargo and easily justified the expense of its construction.

In my opinion, the engineering associated with the 1875 hydraulic lift was ahead of its time. It is apparent that the construction materials and paint systems available at the time were unable to withstand the highly corrosive atmospheric conditions which existed at Anderton at the end of 19th century. This is, to some extent demonstrated by the fact that Lift No. 1 on the Canal du Centre in Belgium, which was built in 1888 also to a design by Edwin Clark, is still in operation today. It was the vulnerability of the construction materials of the 1875 lift at Anderton that led to its early demise in 1908.

In contrast, the structure which was erected over the original lift as part of the electrical conversion, is impressive, if for nothing else, for its sheer bulk. It would be difficult to argue against it being labelled as '*the country's ugliest structure*', especially when viewed in its current denuded and rusted condition. The worth of the 1908 electrical lift is not so much the nuts and bolts of the physical structure, but in the manner of its construction; it was erected and commissioned with minimal disruption to the functioning of the hydraulic lift which it was being built to replace, thereby avoiding long and expensive shutdowns and the consequential risk of driving valuable traffic onto other routes.

The electrical conversion was another brave move by the Trustees in the face of adversity but once again their courage was rewarded by a structure which turned out to be reliable and efficient. Similar to the hydraulic lift, the converted lift proved to be commercially very successful and remained in operation for nominally 75 years. By the time of nationalisation in 1949, it had conveyed more than 4 million tons of goods.

Now all we can do is wait for British Waterways' work to restore the Anderton Boat Lift to be completed. And so, to Jack Horton's and many others' pleasure, the Anderton Boat Lift will, one or two years into the third millennium, once again be conveying boats through the vertical distance of 50 feet 4 inches between the Weaver Navigation and the Trent & Mersey Canal.

★ ★ ★ ★ ★ ★ ★ ★ ★ ★

Finally, at the same time as preparing to restore the Anderton Boat Lift, British Waterways are developing plans for a new, futuristic looking, boat lift in Scotland. The new structure, referred to as the Falkirk Wheel, is to replace a flight of 19th century locks between the Forth & Clyde Canal and the Union Canal. If progressed, it will be the first boat lift to be built in Britain since the construction of the Anderton Boat Lift in 1875. The Falkirk Wheel will have a capacity of 600 tons, or 300 tons per tank, and will be capable of transferring boat vertically over a distance of nominally 100 feet in 15 minutes; effectively twice the size of the Anderton Boat Lift. The total cost of the project is predicted to be £17 million, which includes an aqueduct, a new canal link, a basin and other associated works.

It will be interesting to see if the Falkirk Wheel, when completed, attracts as many plaudits as the Anderton Boat Lift deservedly attracted when it was first opened.

INFORMATION SOURCES FOR CHAPTER 11

Article in The Times *newspaper, 31 March 1999.*

Resubmission of Heritage Lottery Fund Bid, *by British Waterways and the Anderton Boat Lift Trust, April 1998.*

Anderton Boat Lift Restoration – Study for 1875 Restoration and 1908 Static Monument, *by Lewin, Fryer & Partners, April 1998.*

The Falkirk Wheel – Facts and Figures. *Information leaflet published by British Waterways.*

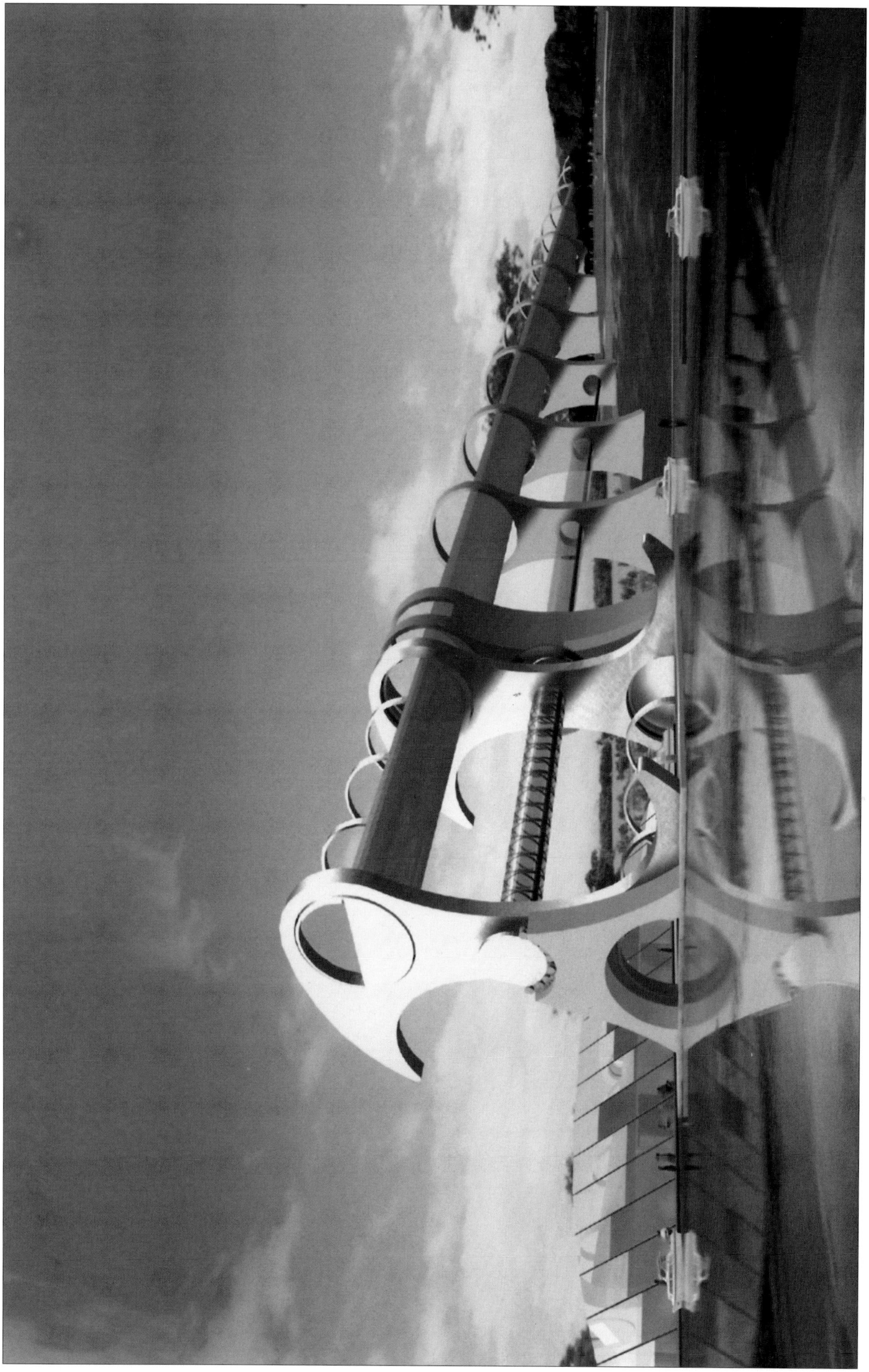

A computer generated image of the proposed Falkirk Wheel in Scotland. It will link the Forth & Clyde Canal with the Union Canal, some 100 feet above. As the lift slowly rotates about the central axis, the two cradles trotate in the opposite direction in order to stay level. The central axis also houses an enclosed viewing platform and there is also a visitor centre, shown behind on the left. Simple in concept but futuristic and stunning in design, it will be certain to generate a huge amount of publicity for our waterway network.

Courtesy British Waterways

INDEX

(NB: names of vessels and publications are italicised)

A view inside the operating cabin in the 1930s, showing the controls for the western caisson.
Courtesy The Boat Museum, Ellesmere Port